Basic Manufacturing

To Diane, Deborah and Angela Saunders, Margaret Cunnew and Caroline Nottage

Basic Manufacturing

Ken Saunders, Peter Birt, Michael Brian,
Terry Cunnew

Revised by Roger Timings

OXFORD AUCKLAND BOSTON JOHANNESBURG MELBOURNE NEW DELHI

Butterworth-Heinemann
Linacre House, Jordan Hill, Oxford OX2 8DP
225 Wildwood Avenue, Woburn, MA 01801-2041
A division of Reed Educational and Professional Publishing Ltd

ℛ A member of the Reed Elsevier plc group

First published 1999

British Library Cataloguing in Publication Data
Basic manufacturing
    1. Production engineering
    I. Timings, R. L. (Roger Leslie), 1927–670

Library of Congress Cataloguing in Publishing Data
Basic manufacturing   /   edited by Roger Timings ... [et al.].
      p.   cm.
    ISBN 0 7506 3650 5
    1. Production engineering.   2. Manufacturing processes.
    I. Timings, R. L. (Roger Leslie)
    TS176.B37   1999
    658.5--dc21
                                                                99–20869
                                                                CIP

ISBN 0 7506 3650 5

Typeset in 10/12pt Palatino by Laser Words, Madras, India
Printed in Great Britain

Contents

Preface

This book has been written to satisfy the needs of students following the GNVQ Manufacturing (Intermediate) syllabus of the Business and Technology Education Council (BTEC). The book is divided into four chapters. Each chapter covers a complete BTEC Unit. Thus Chapter 1 covers all the syllabus requirements of BTEC Unit 1. BTEC Units are subdivided into Elements. Similarly, each chapter of this book is subdivided into main sections called Elements and is numbered accordingly, for example Element 4.3 in Chapter 4 corresponds to Element 3 in Unit 4. The Elements are further broken down into Sections and numbered accordingly. Thus Section 4.3.2 is the second Section in Element 3 of Chapter 4.

Each chapter is prefaced by a summary of its contents. As you work through the chapter you will come across panels of self-assessment question at the end of each main topic area. For example, 'Test your knowledge 3.2' is the second set of self-assessment questions in Chapter 3. At the end of each Element is a summary entitled 'Key Facts'; these briefly review all the main topic areas covered in the main body of the Element. Finally, each Element finishes with suggestions for Evidence Indicators, these are based on the recommendations of the BTEC publication Mandatory and Core Skill Units.

Every effort has been made to keep the writing style friendly and interesting. To avoid long and tedious descriptions, bullet pointed summaries are used wherever possible, together with illustrations, flow charts and tables.

Roger Timings

The world of manufacturing

| Summary | This chapter is concerned with Unit 1 of the GNVQ Manufacturing (intermediate). The *unit* is subdivided into four *elements*. The first element (1.1) describes the key features of the main United Kingdom (UK) manufacturing sectors, the types of product they produce and their relative importance to the overall economy of the UK. That is, the role of our manufacturing industries in national wealth creation. The second element (1.2) investigates production systems from 'jobbing' workshops making one-off prototypes to large-scale mass production. The third element (1.3) investigates and describes manufacturing organization. Finally, the fourth element (1.4) identifies the environmental effects of production (manufacturing) processes in terms of pollution, waste disposal and energy used. |

1.1 The importance of manufacturing to the economy of the United Kingdom

1.1.1 What is manufacturing?

Manufacturing can be defined as *the conversion of raw materials into useful articles by means of physical labour or the use of power driven machinery.*

In prehistoric times, cave dwellers found that if a lump of flint was struck with another stone the flint could be turned into a sharp spearhead or arrowhead. They took the flint as their raw material and, by means of physical labour, converted it into something useful. They were manufacturing. By converting a useless piece of flint into a useful tool, they made it easier to defend themselves and to hunt for food in order to provide the tribe with its next meal.

Manufacturing has come a long way since then. Today, although there are still many skilled craftsmen who make things by hand, the manufacturing industries which are of significant economic importance to the nation are usually large concerns employing hundreds and in some cases thousands of people. They need large buildings, complex machinery and major investments of capital. They increasingly market and sell their products not just in one country but around the world.

Manufacturing as we understand it today began in what is called the *industrial revolution*. The UK was the first country to undergo the change from a largely agricultural economy to full-scale industrialization. The foundations for this industrial revolution were laid in the seventeenth century by the expansion of trade, the accumulation of wealth and social and political change. This was followed in the eighteenth century by a period of great discoveries and inventions in the fields of materials, transportation (better roads, canals and railways), power sources (steam) and in the mechanization of production.

The momentum for change originated with the mechanization of the textile industries (spinning and weaving) which, in turn, resulted in these cottage industries being replaced by the factory system. This increased the need for machines and the power units that would drive these machines more reliably than water wheels which came to a halt during periods of drought. The mechanical engineering industry was born out of the demand for machines and motive power sources of ever increasing size and sophistication. Fortunately, at this time, a series of inventors and engineers in the UK developed the early steam pumping engines into 'rotative' engines suitable for driving machines. The transportation of raw materials and finished goods was transformed by the development of the railways and by steam powered boats. At last, the manufacturing industries and the transportation of their goods were free from the vagaries of the weather and the limitations of horse drawn vehicles.

Mechanization and reliable power sources enabled UK manufacturers to turn out more goods than the UK alone needed. These surplus goods were sold abroad, pointing the way to the worldwide trade in all sorts of manufactured goods which we take for granted today. The ships taking the products manufactured in the UK to other countries brought raw materials from those countries back to the UK to sustain further manufacture.

Manufacturing is a commercial activity and exists for two purposes:

- *To create wealth.* There is no point in investing your money or other people's money in a manufacturing plant unless the return on the investment is substantially better than the interest that your money could earn in a savings account.
- *To satisfy a demand.* There is no point in manufacturing a product for which there is no market. Even if there is a market (a demand), there is no point in manufacturing a product to satisfy that demand unless that product can be sold at a profit.

We have already defined manufacturing as the conversion of raw materials into useful articles by means of physical labour or the use of power driven machinery. When this conversion takes place there is *value added* to the raw materials. This increase in value represents the creation of wealth for both the owners of the manufacturing enterprise itself and the nation as a whole. Manufacturing makes a vital contribution to the local and national economies of the UK. Currently it is estimated that 4.3 million people representing over 20% of all employed persons in the UK are engaged in the manufacturing industries. The survival of many local economies depends solely upon manufacturing in its widest sense and manufactured goods account for roughly 80% of all UK exports and 20% of the Gross Domestic Product (GDP). Gross Domestic Product and Gross National Product (GNP) will be discussed in Section 1.1.8.

Test your knowledge 1.1

1. Describe briefly what is meant by 'manufacture'.

2. State the main purposes of manufacture.

1.1.2 The main UK manufacturing sectors

We have already seen that the manufacturing industry is vital to the UK economy. Without the income generated by manufacturing and the hundreds of thousands of jobs which the manufacturing industry creates, the country would be unable to sustain an acceptable level of employment and pay its way in the world.

Manufacturing takes many different forms. To make the industry easier to study, we will break it down into six main *sectors* which we will then examine individually. These sectors are shown in Fig. 1.1.

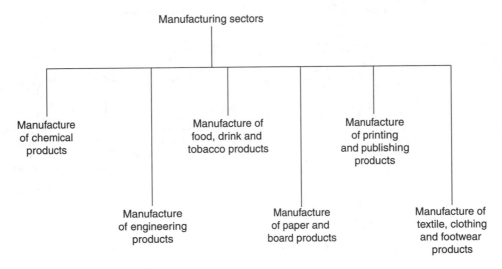

Figure 1.1 Main UK manufacturing sectors.

Let's now look at each of these sectors in turn in order to find out which are the leading companies in each sector and what it is that they produce.

1.1.3 The chemical industry

Chemical manufacturing covers a range of different activities, starting with the manufacturing of the basic chemicals (feed stocks) from raw materials. These feedstocks are then used by other parts (subsectors) of the industry in more complicated processes. Chemicals are defined as being either *organic* or *inorganic*.

Organic chemicals consist of complex *carbon compounds*. Carbon forms more compounds than any other chemical. It forms the complex long chain molecules found in animal and plant life (hence the name *organic* chemicals). Carbon also forms the long chain molecules found in the byproducts of the fossil fuel industries which are widely used in the manufacture of plastic materials. We ourselves are built from carbon chains. Some organic chemical substances upon which we depend are shown in Fig. 1.2. As you will see, many organic chemicals are naturally occurring substances. Other organic chemicals are synthetic and are manufactured from a variety of raw materials.

Inorganic chemicals are all the substances that are *not* based upon complex carbon chain compounds–for example, chemicals made from mineral deposits such as limestone, rock salt, sulphur, etc., as shown in Fig. 1.3.

Both organic and inorganic chemicals are manufactured and used in very large quantities, and this is reflected in the size of the plants in which they

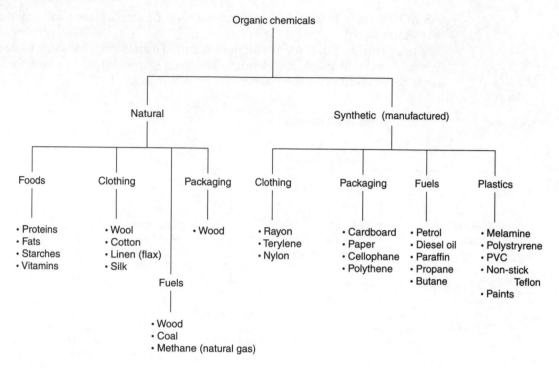

Figure 1.2 Some organic chemicals.

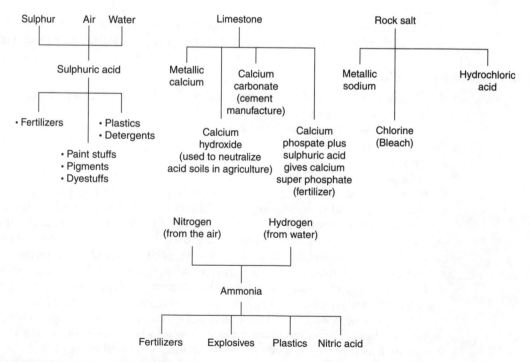

Figure 1.3 Some inorganic chemicals.

are made. In addition, many chemicals, if they are not properly handled, can be extremely dangerous. The size of the factories, the complexity of the manufacturing processes and the need for stringent safety precautions means that the whole business is very expensive, with the result that there are only a few, very large companies which dominate the sector. Some of these are listed in Table 1.1. Note that in this table, and in the following tables, the companies are listed in order of the capital funds employed.

Table 1.1 Top 5 UK chemical companies

No.	Company	Capital employed £000	Turnover £000	Number of employees
1.	Imperial Chemical Industries plc	6 023 000	10 520 000	64 000
2.	The BOC Group plc	3 074 700	3 752 100	40 495
3.	Laporte plc	654 800	1 063 400	6 979
4.	Albright & Wilson plc	409 100	703 400	4 035
5.	Allied Colloids Group plc	259 599	393 390	3 059

Source: The Times 1000 (Times Books) 1998.

You can see that in the UK the chemical industry is dominated by Imperial Chemical Industries plc both in respect of turnover and in the number of persons employed. In terms of turnover it represents over a quarter of the total UK chemical manufacturing industry. The companies listed make products such as plastics, fertilizers, paints, adhesives, explosives, synthetic fibres, and provide gases such as oxygen, acetylene, nitrogen, argon and carbon dioxide for industry, and gases such as oxygen and anaesthetics for hospitals. Other chemical manufacturing companies specialize in domestic and health care products. These range from medicines and disinfectants to cleaning materials and polishes. Some typical companies in this field are listed in Table 1.2.

Table 1.2 Top 5 UK health and household companies

No.	Company	Capital employed £000	Turnover £000	Number of employees
1.	Smithkline Beecham plc	4 783 000	7 925 000	52 900
2.	Glaxo Wellcome plc	4 027 000	8 341 000	53 808
3.	ZENECA Group plc	3 108 000	5 363 000	31 100
4.	Reckitt & Colman plc	1 616 800	2 278 385	17 425
5.	Rentokil Initial plc	620 300	2 269 600	107 767

Source: The Times 1000 (Times Books) 1998.

Another factor which ensures that the chemicals manufacturing sector is dominated by a handful of large operators is that they have to put a lot of money every year into research. Sometimes that research will come to nothing and the money invested will be lost. Sometimes the research will be successful and will be developed into a product that will generate a lot of income. Only very large companies can afford to take this sort of gamble. The profits from the successful research not only has to pay for the money lost on unsuccessful research, it also has to fund the next generation of research.

Many manufactured chemical products play an important part in everybody's daily life. You can use a detergent to wash up after cooking a meal or you can use it to clean your car. Dyes give colour to your clothes, your wallpaper and your carpets. Fertilizers ensure that there is enough food in the shops at a price you can afford.

Without paints all the woodwork in your home would soon decay and rot, while your car or motorbike would quickly become rusty. Pharmaceuticals–medicines–can keep us healthy and, in extreme cases, save our lives.

Polymers (the so-called plastic materials) are used for clothes, ropes, electrical insulators, sports gears and a host of other products upon which modern society depends. Without plastics there would be no CDs, no computers, not even the old-fashioned records made from vinyl pressings, and there would be no tape from which audio and video recording media are made. Without printing ink you wouldn't be able to read the sports pages in your newspaper because there wouldn't be a newspaper.

You may think that, as a law abiding citizen, explosives play no part in your life. However, explosives are widely used in mining and quarrying for the raw materials which are processed to produce things which we use everyday–for example, bauxite is mined to be processed into aluminium and there are few households which do not own at least one aluminium saucepan.

As you can see, the chemicals industry plays a part in nearly every aspect of our daily lives, from the clothes we wear to the games we play and the cars we drive. Life would be very different without them.

Test your knowledge 1.2

1. List the main groups of chemical products manufactured in the UK.

2. Imperial Chemical Industries plc is the largest UK manufacturer of chemical products in the UK. List the main product groups of this company.

3. The BOC Group plc manufactures and supplies industrial gases. Find out which of these gases are extracted from atmospheric air and briefly explain how this is done.

4. Name the products associated with the following companies:

   (a) Glaxo Holdings plc;
   (b) Fisons plc;
   (c) ZENECA Group plc;
   (d) Procter and Gamble plc.

1.1.4 The engineering industry

The engineering industry is very diverse and we need to divide it up into a number of subsectors in order to simplify our study of it. The main subsectors are shown in Fig. 1.4. All these subsectors use metal and plastic raw materials.

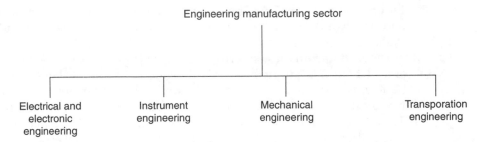

Figure 1.4 Engineering manufacture.

Some use naturally occurring substances such as wood for pattern-making for the metal casting industry which also uses sand for the moulds into which the molten metal may be poured. Natural rubber obtained from the sap of the *Hevea brasiliensis* tree is used for making road vehicle and aircraft tyres. Glass is used in the manufacture of optical instruments and as fibres for reinforcing plastics. The main raw materials for making glass are silica sand, soda ash (crude sodium carbonate) and quick lime (obtained from limestone). Various metal oxides may be added depending upon the properties required. These various substances are mixed together and melted in furnaces to make glass.

Before we start to look at the various product manufacturing companies, we need to consider the sourcing of the raw materials. We have already seen that plastic materials are produced by the chemical industry for manipulation by spinning, weaving, moulding and fabrication into finished products by other industries. As stated above, glass is also produced by chemical reactions between various substances in the glass making 'tank' furnace. Metals are also extracted from mineral ores by chemical reactions. Other physical/chemical processes are used to refine the raw material into useful materials for the engineering industry. The demarcation between chemical engineering and materials engineering quickly becomes blurred.

At the time of the industrial revolution, the UK was largely self-sufficient in metals. In the Black Country–that area surrounding the town of Dudley in the West Midlands–seams of iron ore and coal lay side by side, while limestone lay under Dudley Castle. These are the ingredients needed for the extraction of iron ore in blast furnaces. Tin, copper and zinc were to be found in Cornwall. Unfortunately the demand was great and these resources were small and soon worked out. Iron ore deposits were found elsewhere in the UK but, since the end of World War Two, these have also been worked out.

Metals are still extracted and refined in the UK but almost entirely from imported metallic ores. Some metals such as aluminium and copper have to be refined by electrolytic processes using vast amounts of electrical energy. These are mostly refined where hydroelectric plants can generate electricity cheaply. Except to a limited extent in Scotland, there are no hydroelectric schemes in the UK. The continental countries of Europe are more fortunate in this respect.

Iron and steel manufacture dominates the metal manufacturing industry of the UK and, despite having to import the iron ore, we are still one of the largest manufacturers of ferrous metals–metals and alloys based on iron–in

Europe. Because the basic raw material (iron ore) is imported, most iron and steel works are sited on the coasts of Yorkshire, Humberside and South Wales near to large ports.

The manufacture of non-ferrous metals and precious metals (gold, silver and platinum) is, surprisingly, mostly centred in the Midlands despite the need to import the raw materials. In addition, non-ferrous metals are also manufactured on Tyneside and in London, Avonmouth and South Wales. Table 1.3 lists some typical major metal manufacturing companies in the UK and indicates the scale of their contributions to the national economy and the level of employment they sustain. You can see that British Steel is the largest player in this field, while Johnson Matthey specialize in precious metals.

Table 1.3 Typical major UK metal manufacturing and metal forming companies

No.	Company	Capital employed £000	Turnover £000	Number of employees
1.	British Steel plc	6 035 000	7 048 000	50 100
2.	Johnson Matthey plc (precious metals)	547 000	2 528 900	5 624
3.	Glynwed International plc	293 000	1 323 700	13 307
4.	ASW Holdings plc	205 000	531 800	3 514
5.	Murray International Holdings Ltd	109 261	292 635	3 451

Source: The Times 1000 (Times Books) 1998.

Now let's look at those subsectors of the engineering industry that consume the materials we have just considered. They use these materials to manufacture a vast range of products for an equally vast range of markets at home and overseas. These subsectors will be considered in alphabetical order starting with electrical and electronic engineering.

Electrical and electronic engineering

First, let's consider the difference between electrical and electronic engineering.

- *Electrical engineering* is concerned with heavy current devices used in electricity generation and transmission, and in consuming devices such as motors, heaters and lighting.
- *Electronic engineering* is concerned with relatively small current devices such as computers, telecommunications, navigational aids and control systems. It is also concerned with the manufacture of components such as transistors, diodes, capacitors, resistors and integrated circuits for use in such devices.

The electrical engineering subsector not only makes equipment such as generators, transformers and the cables without which no electricity would reach homes and workplaces, it also produces a huge range of products which are complete in themselves and which you are likely to find in the home. Some of these are set out in Fig. 1.5. Domestic appliances are also referred to as 'white goods' and as 'consumer durables'. First because they

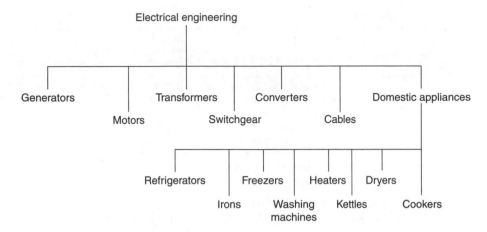

Figure 1.5 Products of the electrical engineering sector.

come in white enamelled cabinets and second because they are used by domestic consumers and the products themselves are durable.

The electronic subsector manufactures products that are used in industry, commerce, service industries and the home. There is often a considerable level of overlap. Desktop computers are equally likely to be found in the home as in offices, banks, hospitals and in manufacturing and design companies. The software will differ but the basic machines are the same. Dedicated computers may be used for machine tool control, in the engine management systems of road vehicles and in navigational aids in aircraft and ships. Navigational aids linked with satellites are now small enough for use in cars and hand-held devices are now available for hikers. Electronic control systems incorporating dedicated computers are used to control domestic devices such as washing machines, toasters and refrigerators. Programmable logic controllers operate the traffic lights at crossroads and can be reprogrammed easily as traffic patterns change. Fig. 1.6 shows a broad breakdown of the products manufactured in this subsector and Table 1.4 lists some typical manufacturers in this sector and indicates their contribution to the UK economy.

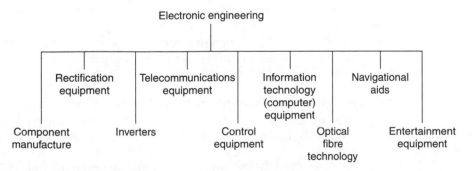

Figure 1.6 Products of the electronic engineering sector.

Table 1.4 Typical Major UK electrical and electronic engineering companies

No.	Company	Capital employed £000	Turnover £000	Number of employees
1.	General Electric Co. plc	4 027 000	6 235 000	82 967
2.	Thorn plc	1 992 800	1 537 400	17 772
3.	Delta plc	506 000	964 257	13 107
4.	Premier Farnell plc	502 200	559 900	4 382
5.	Electrocomponents plc	258 100	792 983	3 440

Source: The Times 1000 (Times Books) 1998.

Instrument engineering

This subsector is concerned with the manufacture of scientific and industrial instruments for measurement, control and diagnosis. However, equipment such as cameras and camcorders are also available on the domestic market. Table 1.5 lists some typical UK companies in this sector and indicates their contribution to the national economy of the UK. These are relatively small companies, since most of the larger instrument manufacturing companies operating in the UK are multinational. Fig. 1.7 shows the general breakdown of the products associated with this subsector.

Table 1.5 Typical UK instrument engineering companies

No.	Company	Capital employed £000	Turnover £000	Number of employees
1.	Vitec Group plc	93 427	148 548	1257
2.	Oxford Instruments plc	83 188	143 576	2411
3.	Halma plc	78 620	173 652	2384
4.	Renishaw plc	66 442	77 077	904

Source: The Times 1000 (Times Books) 1998.

Mechanical engineering

Although classified here as a subsector, mechanical engineering covers such a wide range of engineering activities that it needs to be broken down still further and this is done in Fig. 1.8. Some typical engineering companies in this sector are listed in Table 1.6. Let's now consider the some typical products from each of these groups.

We will consider the groups in alphabetical order starting with agricultural machinery. (Tractors will be considered under *transportation*.) Typical products in each group are shown in Figs 1.9 to 1.16 inclusive.

Transportation equipment manufacturing

Again, this subsector of the engineering industry covers such a wide range of engineering activities that it needs to be broken down still further as shown in Fig. 1.17.

At one time the UK led the world in the manufacture of all types of transportation equipment. However, this lead was already lost early in

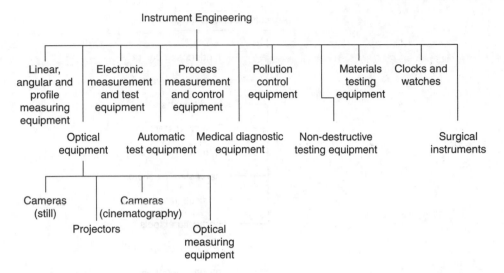

Figure 1.7   Instrument engineering sector.

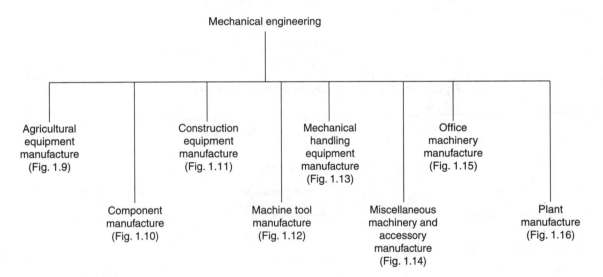

Figure 1.8   Mechanical engineering sector.

Table 1.6   Typical UK mechanical engineering companies

No.	Company	Capital employed £000	Turnover £000	Number of employees
1.	GKN plc	1 521 500	2 873 000	29 515
2.	Turner & Newall plc	994 800	1 956 000	33 893
3.	TI Group plc	772 500	1 552 500	22 650
4.	IMI plc	542 400	1 316 400	16 163
5.	Senior Engineering Group plc	175 770	574 814	6 674

Source: The Times 1000 (Times Books) 1998.

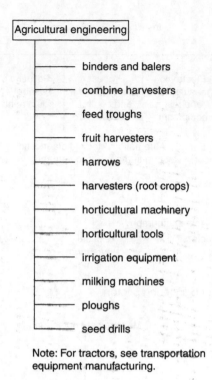

Note: For tractors, see transportation equipment manufacturing.

Figure 1.9 Agricultural equipment manufacturing.

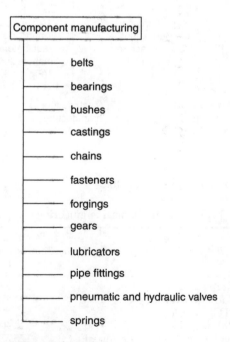

Figure 1.10 Component manufacturing.

Figure 1.11 Construction equipment manufacturing.

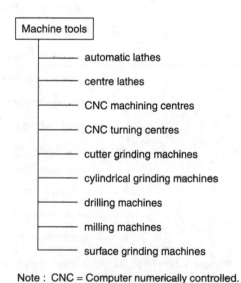

Note : CNC = Computer numerically controlled.

Figure 1.12 Machine tool manufacturing.

this century. From being the largest manufacturer of steam ships we are now the smallest among the maritime nations of Europe. Most of our shipbuilding capacity is kept for strategic purposes to manufacture and maintain vessels for the Royal Navy. Very little merchant shipping is now made and maintained in UK yards. Although we were the first to develop railways and build and equip railways for other countries throughout the world, nowadays the manufacturers of railway locomotives, rolling stock

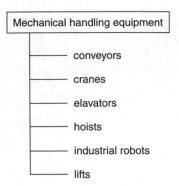

Figure 1.13 Mechanical handling equipment manufacturing.

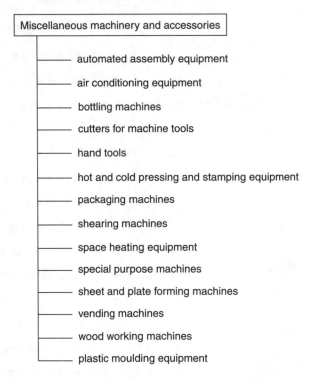

Figure 1.14 Miscellaneous machinery and accessory manufacturing.

and permanent way equipment in the UK only have to satisfy a limited home market with few opportunities for export trade as most countries now have the resources to build their own. Also road transport has reduced the need for railways to a large extent in most countries.

Railways and shipbuilding were intensive users of coal, iron and steel. The rundown of these industries has had a disastrous 'knock-on' effect on iron and steel manufacture and the heavier sectors of the UK manufacturing industry, for example marine engines.

The UK owned car industry has fared little better and, with the sale of Rolls-Royce, no major UK owned car maker is left other than companies

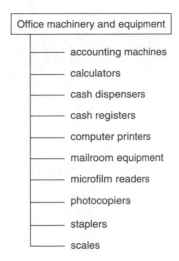

Figure 1.15 Office machinery and equipment manufacturing.

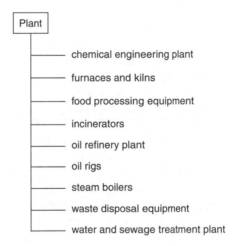

Figure 1.16 Plant manufacturing.

like LTI Carbodies, makers of the famous London Taxi. All the others have either closed or been taken over by multinational companies based overseas. However, these overseas companies manufacturing in the UK–such as the Rover Group owned by BMW, Toyota, Ford (which now owns Jaguar Cars), Nissan, Honda and Peugeot–ensure that the UK is still a major exporter of road vehicles. Further, the UK still makes specialist cars on a small scale such as the Morgan. Also, we are still the world leader for the design, development and manufacture of racing cars. Closely related to the motor industry, *tractors*–mostly for agricultural use–are still manufactured in the UK, as are small and medium sized commercial vehicles. Many tractors and commercial vehicles are exported.

The lead in motor cycle manufacture passed to the Japanese after World War Two. Many famous names of UK companies manufacturing motor

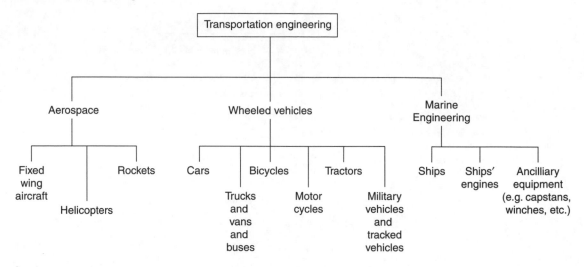

Figure 1.17 Transportation engineering sector.

cycles and pedal cycles disappeared as cars became more available and at a price that ensured popular ownership. However, the pedal cycle industry is expanding again with the advent of the 'mountain' bicycle. This has led to a resurgence in cycling as a healthy recreation and as a means of getting quickly round the crowded roads of our large towns and cities. The use of bicycles also avoids parking problems. There has been some recovery in the manufacture of high performance motorcyles in the UK in recent years.

The aerospace industry is still a major contributor to the transport subsector of the engineering industry. This is largely due to its importance as a manufacturer of defence equipment. The cost of the design and development of modern aircraft means that, outside the USA, new projects demand international cooperation for example, the European Air Bus and the new Euro-fighter. The UK aerospace industry and aviation equipment manufacturing industries are major employers and exporters and are of great importance to the UK economy. Table 1.7 lists some typical companies manufacturing transportation equipment and indicates their contribution to the national economy of the UK.

Table 1.7 Typical transport equipment and vehicle manufacturing companies operating in the UK

No.	Company	Capital employed £000	Turnover £000	Number of employees
1.	British aerospace plc	3 312 200	6 464 000	44 200
2.	Ford Motor Co. Ltd	N/A	5 383	38 400
3.	Rolls-Royce (Aerospace) plc	2 271 000	4 291 000	42 600
4.	Lucas Varity plc	1 620 600	4 009 650	53 100
5.	Nissan Motor Manufacturing (UK) Ltd	N/K	1 054 000	4 974

Source: The Times 1000 (Times Books) 1998.

Although by now you must have thought that the above lists were never ending, nevertheless they are by no means comprehensive and only represent a fraction of range of products manufactured within the various sectors and subsectors of the engineering industry. The overall number of persons working in the various subsectors of the engineering manufacturing industry is shown in Table 1.8. This table also shows the number employed in engineering compared with the total working population of the UK.

Table 1.8 Overall employment provided by the engineering industry in 1996

Sector	No. of employees	% of all manufacturing employees	% of all those in employment
Metal manufacture	110 300	2.8	0.6
Mechanical engineering	482 600	12.3	2.6
Electrical and electronic engineering	433 300	11.1	2.0
Instrument engineering	63 700	1.63	0.3
Transport manufacture	353 800	9.04	1.6
Metal products	1 882 000	48.1	8.7
TOTAL	3 325 700	84.97	15.8

Test your knowledge 1.3

1. List the main subsectors of the engineering industry.

2. Name four major UK companies in each of the subsectors listed in question 1.

3. Find out where the companies named in question 2 are located and suggest reasons for their location.

4. There are many engineered products in your home. Find out:

   (a) how many are produced in the UK;
   (b) the names of the companies by whom they were manufactured;
   (c) the subsector of the engineering industry to which those firms belong.

1.1.5 Food, drink and tobacco

The food, drink and tobacco industries are as diverse as any we have examined so far. Again we need to divide them up into a number of subsectors in order to simplify our study of them. These main subsectors are shown in Fig. 1.18.

Bakery products

A few large companies such as Rank, Hovis McDougal and Allied Bakeries not only dominate the bakery industry, they manufacture flour for both the wholesale and retail markets. The range of products manufactured by the bakery industry is indicated in Fig. 1.19. In addition, there are a number of smaller, independent manufacturers making specialist products such as cakes, pastries and such things as pies and puddings–the last two obviously overlap with those parts of the food industry which provide the fillings. Just as in baking, the majority of biscuits are manufactured by a few large companies–for example, Associated British Foods.

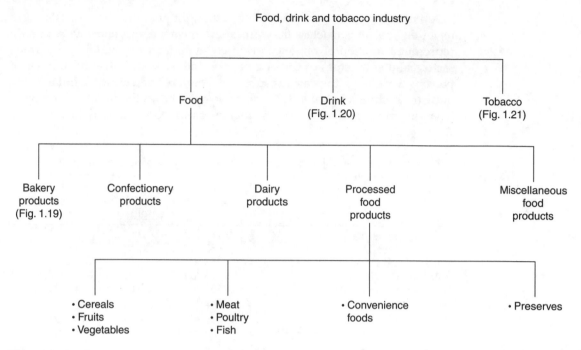

Figure 1.18 Food, drink and tobacco manufacturing sector.

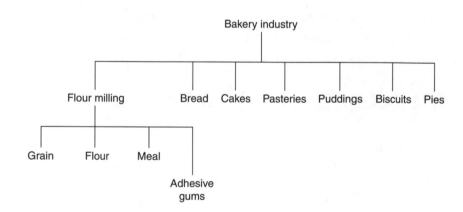

Figure 1.19 Bakery products manufacturing industry.

Confectionery products

The largest manufacturers in this subsector of the food sector are Cadbury Schweppes, Mars and Rowntree Mackintosh, this latter company being part of the international Nestlé Group. They manufacture chocolates, sweets, drinking chocolate, cocoa and chocolate biscuits. Confectionery products also include 'fancy' cakes. A large number of smaller firms manufacture sweets and fancy cakes, although it must be recognized that some, trading under their own brand names, are, nevertheless, owned by the larger groups.

Dairy products

Milk is produced by cows owned by dairy farmers. It is a natural process. A cow only produces milk after calving. So you have to produce calves as well. Some female calves are retained to replace older milking cows when their milk yields fall. These older animals are unsuitable for human consumption but, if free from disease, can form the feedstock for the pet food industry. The majority of calves–male and female–are an important byproduct of the dairy industry and are sold on for fattening for the fresh and processed meat industry. This helps to keep the cost of milk and dairy products at an affordable level. The production of milk and meat is inexorably linked by the laws of nature.

Milk is the bulk raw material of the dairy industry, which this industry then processes. The milk which comes from the farm is not the same milk that is delivered to your doorstep –it is pasteurized and purified beforehand. Pasteurizing is a heat treatment process that kills harmful bacteria and also assists the milk in staying fresh. Some milk is further processed into butter or cheese, yoghurt and cream. Milk from which the cream (fatty content) has been removed is sold as low fat, skimmed or semiskimmed milks. Canned milk is usually either condensed or evaporated to drive off the water content and concentrate the nutrients.

Processed foods (cereals, fruit and vegetables)

Many foods are processed in some way before being made available to the professional chef, cook or the domestic consumer. From the earliest days cereal crops have had to be harvested, threshed to remove the seeds and these seeds or 'ears' have had to be ground to produce flour for cooking. Nowadays we expect our potatoes to be cleaned and prepacked for sale in the supermarkets. In view of the quantities involved the cleaning, grading and packing is a highly mechanized industry. Potatoes are also processed into chips and crisps. The chips are often partly cooked and 'oven ready'. Both these processes are highly mechanized in large and expensive plants that only major companies can afford.

Although many fruits and vegetables are sold fresh when they are in season, soft fruits and vegetables are often frozen so that they are available all the year round. The speed of modern transport and the fact that the seasons alternate between the northern and southern hemispheres of the world means that fresh produce is much more widely available than it used to be. Automated sorting, grading and packaging plants coupled with flash freezing facilities require a large investment and access to mass markets on a global scale, so again food processing is 'big business' involving the largest companies. The same produce is also processed into sauces. Some fruits are canned in syrups to preserve them. This again requires large and expensive automated plants. The engineering industry provides the cans and machinery. *You cannot dissociate the various manufacturing industries; they all depend on each other.*

Processed foods (meat, poultry and fish)

Meat may be sold raw through butchers' shops or it may be bought processed in some form to reduce the time spent in the kitchen in its preparation–for example, ready cooked and canned meats, sausages and frozen poultry. Meats preserved by smoking and curing in salt have long been used for the production of ham and bacon. Although fish is still available fresh, both on the bone and filleted, it is now available frozen, smoked, canned and also reconstituted, as in fish fingers. Byproducts of the fish processing industry are fish blood and bone fertilizers for the horticultural industry.

Processed foods (convenience)

These consist of boxed and frozen, complete, precooked meals containing meat, vegetables and a sauce. They merely have to be reheated in a microwave cooker ready for serving. Vegetarian meals are available as well. Potato crisps and similar prepacked snacks can be included in this category.

Processed foods (preserves)

This sector of the industry includes tinned fruit, vegetables, soups, bottled pickles and sauces, together with preserves such as jams and marmalades.

Miscellaneous food products

Under this heading can be included the production of a wide range of products such as:

Horlicks	Ovaltine	cocoa	drinking chocolate
sugar	table salt	vegetable oils	white fats
margarine	coffee	tea	yeast products
mustard	spices	starch products	pet and farm animal foods

The above are but a few examples of an ever growing range of such products.

The drinks industry

The drinks industry produces all those drinks which – apart from the addition of water in some cases – are ready for immediate consumption. Again, it eases our study of this subsector of the food, drink and tobacco industries to divide it up further as shown in Fig. 1.20.

Soft drinks

These can be grouped as shown above in Fig. 1.20. Let's now look at the individual categories in more detail.

Mineral waters are enjoying an expanding market for both sparkling (carbonated) and still waters. These are filtered and bottled at springs that are free from contamination. The naturally dissolved mineral salts in these waters improve their flavours. Many people find them more palatable than tap water.

Fruit drinks made by crushing fruits and extracting the juice may be produced as frozen concentrates requiring thawing out and the addition of water, or bottled ready for use. When bottled ready for use they are often referred to as *cordials*. They may be pure or they may contain preservatives and added sweeteners. The fruits used are typically oranges, lemons, limes, black currants and unfermented apple juice.

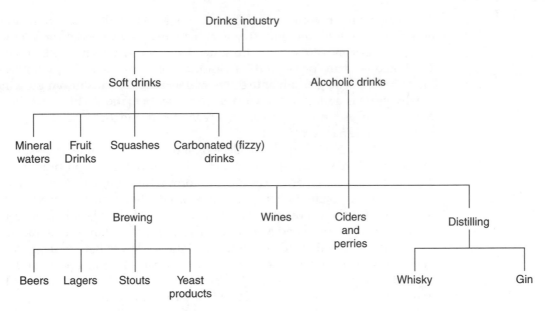

Figure 1.20 Drinks manufacturing industry.

Squashes are manufactured concentrates that require diluting with water to taste. They may contain some fruit concentrates but they contain mainly flavourings, sweeteners, preservatives, added vitamins and colorants. They cannot be considered as 'whole fruit' drinks. The so-called 'slimline' and 'diet' drinks contain artificial sweeteners in place of sugar.

Carbonated (fizzy) drinks include such products as cola, lemonade, tonic water, ginger beer, ginger ale, soda water and others. These may be drunk by themselves or used as *mixers* with wines and spirits. This is slightly confusing because fruit drinks such as tomato juice and grapefruit juice may also be used as mixers. The 'fizz' is produced by adding carbon dioxide gas under high pressure so that it is absorbed into the liquid during processing.

Brewing

By volume, brewing is the biggest sector of the industry, producing alcoholic drinks, providing *beers* (made from grain), *ciders* (made from apples) and, to a smaller degree, *perries* which are made from pears. Beer accounts for the overwhelming bulk of this market, producing draught, bottled and canned beers including bitters, lagers and stouts. The bulk of this production is in the hands of some 10 major brewers and 80 smaller companies.

Brewing can be traced back to the ancient Egyptians. Prior to the fifteenth century most brewing in the British Isles was carried out in the monasteries where the monks produced *ales*. Early in the fifteenth century hops with their preservative and flavouring properties were added and ale became known as 'biere'. This continental spelling was quickly corrupted to 'beer' and this name continues today.

Beer is made from crushed malted barley steeped in boiling water. The liquid (*wort*) is run off for further processing and the residue, known as 'brewers' grains', becomes a byproduct for livestock feedstuffs. Again notice how the waste from one industry is an essential raw material for another. Hops and, in some cases sugar, are added to the liquid which is again boiled. The liquid is again run off and the residual hops become a byproduct used in market gardening as a manure. It is widely used for mushroom growing. Finally yeast is added to the wort to cause fermentation. This is the stage where some of the natural sugars break down and become alcohol and carbon dioxide gas is released.

The carbon dioxide gas is collected and dissolved under pressure in carbonated drinks. So carbonated soft drinks also depend on brewing for their fizz. The yeast reproduces itself to form a frothy mass, most of which is skimmed off. Some is kept for future production and the surplus is a byproduct for the production of yeast foods (such as Marmite) since the yeast is rich in proteins and B vitamins. The liquid now stands for some days to 'mature' and is then filtered and put into bottles or barrels. Lagers are matured for very much longer than ordinary beers. Stouts are made from roasted barley grains sometimes with sugar. The roasting caramelizes the barley to give the stout its distinctive flavour and dark colour.

This topic has deliberately been dealt with in some detail to reinforce how many processes can often generate byproducts that are essential feedstocks for other unrelated industries. It is important to realize that the manipulation of markets, however well intentioned, in one product area can often have serious economic consequences by affecting the steady flow of byproducts that are the essential raw materials for other industries.

Spirit distilling

Drinks with the highest alcohol content are spirits which are manufactured by distillation processes. The main distilled spirits manufactured in the UK are whisky (distilled in Scotland and Ireland), gin and vodka. Other spirits which are popular, but which are produced abroad, are brandy and rum. Scotch whisky is distilled from a malted barley mash and may be sold as pure malt whisky, or a number of whiskies may be mixed and sold as blended whisky. Irish whiskey is distilled from a rye mash. After distillation the spirit is left to mature in oak casks for a number of years. Much of the flavour and the colouring comes from the 'tannin' in the oak. Gin is also distilled from a grain mash but, in this case, it is flavoured with juniper. It may be dry (unsweetened) or slightly sweetened. The alcoholic drinks industry is very heavily taxed and contributes huge sums in revenue to the government.

Wines

The UK also has a small wine industry based mainly on vineyards in the south of England and the Midlands (grapes do not flourish in the colder climates of the more northern parts of the UK). Almost all UK wines are white. Wines are also manufactured on a small scale from imported grapes. As with brewing most of the small independent manufacturers have been absorbed by the larger companies which now control most of the industry. This has come about because of the level of demand and the economics of large-scale production.

Tobacco

The tobacco industry is capital intensive because of the high level of automation involved, particularly in the bulk manufacture of cigarettes. Because of the capital intensive nature of the industry it is controlled by four major companies together with a number of smaller companies which make specialist products for the wealthier smoker. The main groups of tobacco products are shown in Fig. 1.21. Like alcoholic drinks, the tobacco industry is heavily taxed and is a major source of revenue for the government.

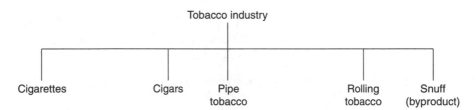

Figure 1.21 Tobacco products manufacturing industry.

Finally we must look at the scale of the food, drinks and tobacco sector of UK industry and its influence on the economy. An indication of this can be gleaned from the details concerning the larger companies involved as shown in Table 1.9.

Table 1.9 Typical UK food, drink and tobacco manufacturing companies

No.	Company	Sector	Capital employed £000	Turnover £000	Number of employees
1.	BAT Industries plc	Tobacco	8 858 000	7 219 000	81 039
2.	Unilever plc	Food	3 165 000	9 814 000	306 000
3.	Whitbread plc	Drink	3 072 000	3 027 200	62 074
4.	Tate & Lyle plc	Food	2 362 500	4 879 000	25 270
5.	United Biscuits (Holdings) plc	Food	853 600	1 986 700	26 100

Source: The Times 1000 (Times Books) 1998

Test your knowledge 1.4

1. Name the main subsectors of the UK food, drink and tobacco industry.

2. Name typical examples of manufactured products from each group that you can find in your home and, where applicable, name the companies.

3. Find out where the companies named in question 2 are located and possible reasons for their location.

4. Write a brief report on the importance to the UK economy of the food, drink and tobacco industry.

1.1.6 Paper and board manufacture

This sector of the manufacturing industry produces the paper for newspapers and books, writing paper, wrapping and packing papers, toilet rolls, kitchen paper and tissues. It also makes boards, boxes, bags and all manner of packaging, as well as such things as wallpaper. The industry divides into two subsectors, one making paper and the other involved in the manufacture of boards (cardboards and mill boards), as shown in Fig. 1.22. Market leaders in this sector are listed in Table 1.10.

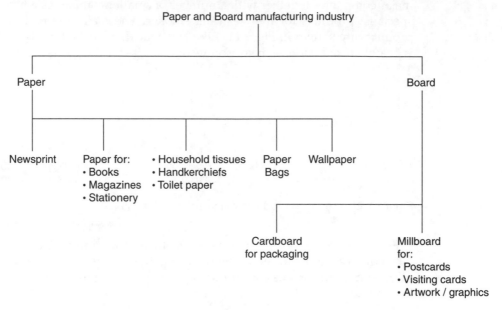

Figure 1.22 Paper and board manufacturing industry.

Table 1.10 Typical UK packaging, paper and printing companies

No.	Company	Capital employed £000	Turnover £000	Number of employees
1.	Arjo Wiggins Appleton plc	1 605 400	3 572 400	19 467
2.	Smith [David S.] Holdings	646 900	1 235 600	10 039
3.	De La Rue plc	322 800	763 800	9 411
4.	Waddington (John) plc	117 969	298 615	3 309
5.	Field Group plc	95 914	201 008	2 289

Source: The Times 1000 (Times Books) 1998

1.1.7 Printing and publishing

Obviously there are areas of common interest between this sector and the paper and board industry, since the latter produces the materials used by the printing and publishing industry. The printing and publishing sector produces not only newspapers, magazines and books but also stationery, official documents, greetings cards, banknotes and postage stamps. The printing industry has seen tremendous changes over recent years, with computerization doing away with many of the skilled printing trades. Mergers in the newspaper, magazine and book publishing sectors have resulted in the industry being controlled by a small number of large companies. Only newspapers are produced 'in-house', magazines, books and other products rely upon the support of a large number of smaller companies specializing in such skilled trades as computerized typesetting, printing and binding. There are a number of general printers who produce stationery and advertising matter for industry and commerce. No detailed figures are included for this sector of manufacturing, which is relatively small compared with those previously discussed, and has little influence on the national economy.

Test your knowledge 1.5

1. Name the main manufacturers of bulk paper and board in the UK, state where they are located and suggest possible reasons that could have influenced their location.

2. Find out and explain the essential differences between publishing and printing.

3. Name the main publishing groups concerned with the production of:

   (a) newspapers;
   (b) magazines;
   (c) books.

1.1.8 Textiles, clothing and footwear

As on previous occasions it is necessary to break down this sector into a number of subsectors as shown in Fig. 1.23. The manufacture of textiles and fabrics requires expensive and large machines, therefore it is centred on a relatively small number of major companies. There are some smaller companies making specialized fabrics that are relatively costly. However, making up the raw materials into garments is still largely labour intensive. This is reflected in Table 1.11, where you can see that although the level of turnover falls rapidly from the leading companies, the number of employees remains surprisingly high. Imported fabrics from the Far East and eastern Europe have seriously challenged the manufacture of textiles in the UK. This has resulted in a substantial reduction in this sector of the manufacturing industry.

The textile industry

Textiles are made by spinning fibres obtained from the wool and fur of animals, or fibres obtained from plants, into yarns which are then dyed and woven into textiles. A number of textiles have the pattern printed onto them after weaving. Some textiles are made from synthetic fibres or blends of natural and synthetic fibres. Let's now look at the main groupings of these products.

Wool

The UK is still one of the world's largest manufacturers of woollen textiles with the industry based mainly in Yorkshire and Scotland. The raw material (fleece) is obtained from the annual shearing of sheep. Sheep grow a heavy coat in the winter for protection against the cold and this is sheared off in the spring so that they are not affected by the heat of the summer. Some of this raw material is obtained from sheep bred in the UK but much is imported, coming largely from Australia and New Zealand. The fleece is spun into yarn and which is then either woven or knitted.

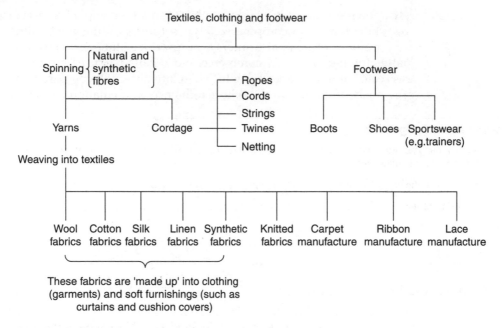

Figure 1.23 Textiles, clothing and footwear manufacturing industries.

Table 1.11 Typical UK textile companies

No.	Company	Capital employed £000	Turnover £000	Number of employees
1.	Coats Viyella plc	1 239 700	2 455 100	69 488
2.	Courtaulds Textiles plc	378 300	1 006 500	21 300
3.	Allied Textile Companies	115 982	289 837	3 272
4.	Readicut International (clothing) plc	88 128	234 280	1 923
5.	Claremont Garments (Holdings) plc	N/A	114 000	3 824

Source: The Times 1000 (Times Books) 1998

Linen

This is woven from fibres from a plant called *flax* which is grown mainly in southern Ireland. However, it is also grown in Northern Ireland and, to a lesser extent, in Scotland. Fine linen products such as tablecloths, handkerchiefs and other household textiles are manufactured in Northern Ireland, while heavy duty linens and canvases as used for tents, tarpaulins and sails are mainly manufactured in Scotland.

Cotton

This does not grown in the UK but raw cotton is imported mainly from America, India and Egypt. The raw cotton is spun and woven into textiles mainly in Lancashire, originally because the damp atmosphere of Lancashire aided the spinning process and helped to prevent the fibres from becoming brittle and breaking. At first, cotton was imported from America and the port of Liverpool became the major west coast port servicing this transatlantic trade. The cotton industry has suffered in recent years from the growth of textile manufacturing in the Far East and the importation of cheap cotton textiles. As previously stated, cotton is widely grown in the Middle East and the Far East where there is abundant cheap labour to process it into textiles that are more profitable to export than the raw cotton.

Synthetic fibres

The manufacture of synthetic fibres is a branch of the plastics industry. The plant is capital intensive and the industry of synthetic fibre manufacture is dominated by a few large companies. The fibres are manufactured from rayon (cellulose base), nylon and polyester (terylene). These fibres are spun into yarns which are then woven into textiles. One of the problems of synthetic fibres is that they are largely non-absorbent and, therefore, difficult to dye. Further, although strong, they tend to fray when subject to abrasive wear. Frequently, textiles are made from mixtures of cotton or wool with synthetic fibres as these are more pleasant when worn next to the skin.

Carpet manufacture

The UK also has a substantial carpet making industry, including famous names such as Wilton and Axminster which have an international appeal and reputation. These high quality carpets are woven and have a cut pile. Cheaper carpets use synthetic fibres for the pile or mixtures of synthetic and natural fibres. Kidderminster in the Midlands was also a major carpet manufacturing area but it has suffered from cheap foreign imports in recent years. The foundation or backing of carpets is largely made from the jute fibre.

Jute manufacture

Unlike cotton fibres which are obtained from the seed pod of the plant, jute is made from fibres found in the stems of the *Corchorus* plant grown mainly in Pakistan. It is spun and woven into coarse cloth in both Pakistan and India. Raw jute is imported into the UK where it is also manufactured into yarns and hessian fabric. As previously stated, jute is widely used as the foundation or backing for carpets and rugs. A coarse cloth called *hessian* is made from jute and this is used for making sacks. Hessian is also used in the manufacture of upholstery.

Miscellaneous

Not all textile yarns are made up into cloths and fabrics. Some are converted into lace, ribbons and cordage used in the manufacture of strings, twines, nettings and ropes. The UK has one of the largest cordage manufacturing industries in Europe.

The clothing industry

Most of the textiles produced in the UK are used in the making of clothes. The UK has one of the largest clothing manufacturing industries in Europe. Unlike many other sectors of UK industry previously discussed, the clothing sector is spread over a very large number of small concerns. UK clothing and fashion designers have an international reputation for excellence. The industry is labour intensive.

- The clothes are cut out by manually controlled, power driven shearing machines.
- They are then made up on power driven sewing machines operating very much faster than their domestic counterparts.
- Finally the garments are pressed and packaged.

Computerization is starting to creep into the larger concerns, especially for tailoring where computer numerically controlled (CNC) laser cutting out machines are being used. This enables an individual customer's measurements to be entered into the computer and the suit is cut out exactly to size ready for making up.

Knitwear is largely manufactured in Scotland and in the East Midlands. Handmade knitwear is largely made in the Scottish islands as a 'cottage industry'. Mass produced knitwear is made in and around the cities of Leicester and Nottingham. Unlike woven cloths, the yarn in knitwear is made directly into the finished garment ready for wear. There is no cutting out and no waste. Typical products are underwear, socks, stockings, tights, gloves, jumpers, pullovers and cardigans. Hats, gloves and fur goods (natural and imitation) are also made in the UK but in relatively small quantities by specialist firms.

Fabric industry

Laser cutting machines are gradually being replaced by high power, water jet cutting as this does not burn the fabric.

Footwear

The footwear industry in the UK has suffered in recent years from the importation of cheap products. Leather has largely given way to synthetic materials and composites. Fashion products tend to be made in Italy and sportswear is largely imported from low cost regions. Nevertheless quality footwear is still made in the UK and there are also small firms which make boots and shoes to individual order.

Test your knowledge 1.6

1. Name the main types of fabrics and textiles manufactured in the UK and name the major companies producing them.

2. Examine your own clothes.

   (a) From what sort of textiles are they made?
   (b) Try and find out the names of the companies that made up the clothes.
   (c) What size firms are they and where are they located?

3. Name the main boot and shoe manufacturers in the UK and find out the range of products they make and how many people they employ.

4. Name the main manufacturing regions in the UK for textiles, clothes and footwear and suggest reasons that could have influenced their location.

1.1.9 The importance of the main manufacturing sectors to the UK economy

So far we have looked at the main manufacturing sectors of the UK in terms of the range of industries that fall within this heading, the number of persons employed and the volume of the business conducted in terms of turnover. However, large turnovers do not represent large wealth creation.

Suppose you manufacture bird tables in your garage as a paying hobby. At the end of the year you can sit back in pride for you have employed and paid several members of your family for their help, you have produced a thousand bird tables selling at £12.00 each and you can show a turnover of £12 000.00. Unfortunately each bird table cost you £15 to make, so you made an overall loss of £3000.00. Not much wealth creation here; in fact you are worse off than when you started despite that impressive turnover.

To be so successful in business, the companies listed in the various tables earlier in this element make a profit and create wealth for themselves and for the nation. However, you have to look beyond the turnover when assessing the contribution made by any manufacturing concern or sector to the national wealth.

You may also wonder how the manufacturing sectors can contribute to the national wealth when so many of them are multinationals based outside the UK and when so many UK companies have branches abroad. Let's look at this more closely. Two financial terms already mentioned and which you may have seen used in the newspapers and on television are:

- Gross Domestic Product (GDP).
- Gross National Product (GNP).

Gross domestic product (GDP)

This is the total value of all the output produced within the borders of the UK. So GDP is defined as the value of the output of all resources situated within the physical boundaries of the UK no matter where the owners of those resources live. Therefore the UK branch of a multinational company based abroad contributes to the GDP. On the other hand, the wealth created by the output of an overseas branch of a UK company is not included in the GDP. Further, outputs in this context not only include manufactured products (*visible earnings*) but services such as banking and insurance (*invisible earnings*) as well. In many ways, the GDP is more important in assessing the economic health of the nation than the GNP, to be considered next.

Gross national product (GNP)

This is the total value of all the goods and services produced by companies owned by UK residents and UK financial institutions. These UK owned companies not only lie within the physical borders of the UK but overseas as well – for example, the wealth created by the overseas branch of a UK based company would count towards the GNP. On the other hand, unlike the GDP, wealth created within the UK by companies owned by foreign interests does not count towards the GNP. Therefore the GNP is defined as the value of the total outputs of all the resources owned by citizens of the UK wherever the resources themselves may be situated. Again, outputs include the invisible earnings of the service industries as well as manufactured goods.

Balance of trade

The UK is not self-sufficient in either food, raw materials or manufactured goods. In recent years the manufacturing base of the UK has shrunk as a result of firms being forced out of business by cheaper foreign imports. This is serious because we have to be able to pay our way in the world. To remain financially sound we must ensure that our income exceeds our expenditure. This applies both to our personal domestic lives and to the nation as a whole. On a national basis the total income we earn from abroad for our exports must 'balance' or exceed the total payments we must make abroad to pay for our imports. This 'balance' is called the *balance of trade*.

With the shrinkage of its manufacturing base, the UK balance of *visible* exports (e.g. manufactured goods, foodstuffs and fossil fuels) is, generally, insufficient to pay for our imports. Fortunately we have a relatively healthy sale of services abroad (e.g. banking, insurance and consultancy) and these *invisible* exports usually give us a trading balance, as does inward investment when foreign companies build factories in the UK. When you go abroad on a holiday you take money out of the country and the cost of your holiday counts as an import. When tourists come to this country they bring money into the country. Therefore in-tourism counts as an export. The tourist industry is a vitally important invisible earner for the nation.

Let's now consider the contribution made by the manufacturing industries to the UK's balance of payments. We can obtain information on the UK's balance of payments from the 'Pink Book' published by the Office for National Statistics entitled *United Kingdom Balance of Payments*. Table 1.12 shows some typical examples for various sectors of the manufacturing industry and is derived from statistics for *Trade in Goods by Commodity* for the year 1996 as published in 1997.

Table 1.12 Trade in goods by commodity for 1996 (typical examples)

Sector	Imports (£ millions)	% of total UK commodity imports	Exports (£ millions)	% of total UK commodity exports
Food, drink and tobacco	16 644	9.3	11 296	6.7
Chemicals	18 047	10.1	22 202	13.4
Motor cars	11 083	6.2	8 520	5.1
Consumer goods	20 523	11.5	15 303	9.2
Ships and aircraft	3 915	2.2	5 510	3.3
Crude oil	3 812	2.1	7 526	4.5
Basic raw materials (including iron and steel)	6 528	3.7	2 804	1.7

Total commodity imports £(M) 178 938
Total commodity exports £(M) 166 340.

The overall figures for 1996 show that:

- we imported £(000) 178 938 000 worth of goods;
- we exported £(000) 166 340 000 worth of goods;
- there was an overall trading loss on the UK's balance of payments for manufactured commodities of £(000) 12 598 000.

Fortunately the Services Sector made a trading profit of £(000) 8 247 000 and investment income for the same period amounted to £(000) 9 652 000. Therefore for 1996 the UK's balance of payments was in credit.

Test your knowledge 1.7

1. For each of the commodity sectors in Table 1.12, state whether the balance of trade was in favour or not in favour of the UK and calculate the actual amount in £ millions as a profit or loss.

2. Calculate the overall balance for this sector, stating whether it shows a profit or loss for the national balance of payments.

It would appear from the above statistics that the manufacturing industries of the UK have much to answer for since, overall, we import more manufactured goods than we export. In fact you may wonder why we persist in maintaining a manufacturing industry at all. However, care must always be exercised when interpreting statistics.

- Not all the products imported in any one sector are raw materials for manufacturing. Many are finished products sold directly to the consumer. When the cost of these is stripped out of the equation the balance in favour of manufacturing becomes much more favourable.
- If we reduced or closed down the manufacturing industries of this country, then many more imported goods would be required to satisfy the national domestic market. The more of these goods we can manufacture in the UK the fewer we have to import. Further, the cost of imported raw materials is always lower than the cost of finished products.
- Finally, we must remember that manufacturing employs a great many people in the UK. Any reduction in manufacturing would increase unemployment. Only persons who are employed can create wealth for the nation. Remember the statistics with which this element was opened. Currently it is estimated that the 4.3 million people representing over 20% of all employed persons are engaged in the manufacturing industries, and that manufactured goods accounted for roughly 80% of all UK exports and 20% of the Gross Domestic Product (GDP).
- Therefore it should be the aim of the nation to expand rather than contract its manufacturing base so that more citizens of the UK can be employed in wealth creation and contribute to a sound national economy and balance of payments.

Test your knowledge 1.8

1. Write a brief report on the importance of the manufacturing industries to the UK economy in terms of:

(a) wealth creation;
(b) balance of trade;
(c) the beneficial effects of manufacturing on society.

1.1.10 Industry and its location

Many factors have influenced the location of industry in the UK over the centuries and from time to time these factors have changed as a result of the changing needs of the population, as a result of technological advance and as a result of political expediency. It must not be thought that manufacturing only commenced with the industrial revolution. The main changes that came about during the eighteenth and nineteenth centuries were:

- Manufacturing processes using hand skills were progressively replaced by mechanized manufacturing.
- Bulk manufacture by large and costly machines could only be carried on in factories instead of in the homes of the skilled workers of the previous era (cottage industry). Hence the rapid growth of the mill towns of northern England and the migration of labour from the surrounding rural areas into such towns.
- Transportation which was improved by the road and canal builders and, later, by the railways. This enabled materials and finished goods to be moved in bulk. They no longer had to be produced locally.

Prior to the industrial revolution, transportation relied upon horse drawn vehicles and pack horses. The loads that could be moved were limited and, in the winter, many roads–little better than tracks–were impassable. Such manufacturing as there was had to be carried on in the homes of craftspersons–usually by hand–adjacent to sources of raw material and within easy reach of centres of population. What little bulk transport there was relied upon rivers such as the Thames and the Severn, together with coastal trading ships. Thus towns tended to grow in size only where access by water was available.

Many towns had already grown up as a result of being adjacent to the medieval religious centres (cathedral cities) and/or where a Royal Charter had been granted to hold a market, that is, the creation of market towns. Mostly trades people and farmers sold and bought locally grown produce, livestock and locally manufactured goods at such markets, but itinerant (travelling) craftspersons and traders also brought their wares if these were small and/or light and could be readily transported.

That the industrial revolution took root in the Midlands and the North resulted from a number of factors:

- The cotton industry was already established in Lancashire because of the damp climate that aided cotton spinning and access to the transatlantic shipping trade through the port of Liverpool at a time when most raw cotton came from the cotton fields of the USA.
- The woollen industry grew up in Yorkshire using the fleeces from the sheep farmed in the Yorkshire dales.

From this background men like James Hargreaves developed his spinning jenny that improved and speeded up the spinning of cotton fibres into thread; Richard Arkwright who improved on this process with his spinning frame that combined the processes of carding, drawing and roving with spinning; Samuel Crompton whose spinning mule could spin stronger threads that could be used for the warp as well as the weft; and Edmund Cartwright who developed the first power driven loom for weaving cloth.

The engineering industry grew up from the combined skills of the village blacksmith and the finer skills of clock and instrument makers. The early machine tool industry grew up initially among the skilled clock and instrument makers of London. However, the limited transport facilities of this age (horse and cart or horse drawn narrow boats on the canals) meant that many new and heavier engineering firms had to be located near to their raw materials of iron and coal, and to their customers, the mill owners of Lancashire, Yorkshire and Nottinghamshire. Thus young and ambitious, highly skilled mechanics migrated from London, where they had learned their trade, to the Midlands and the North.

At first the newly mechanized manufacturing processes had to be driven by water power. This was not reliable and subject to climatic changes. To overcome the problems associated with water power, Matthew Boulton, a major and influential Birmingham manufacturer, set up the young Scotsman, James Watt, in business to manufacture his rotative steam engines to drive the new machines. With this, James Watt was able to draw on the iron producing industry that was growing up in the so-called Black Country around Dudley and Tipton where coal, iron ore and limestone was available.

Another factor often overlooked is that of religious persecution. Many of the wealthy manufacturing families of the industrial revolution followed the newer non-conformist religions (e.g. the Methodists, the Baptists and the Quakers) rather than the established Church of England. Such 'non-conformists' were barred from UK universities and the professions until the 1870s. They were also prohibited from trading in the old, incorporated towns and cities where non-conformity was punishable by heavy fines and even imprisonment. Therefore they set up their businesses in the newer towns of the Midlands and the North, in towns such as Birmingham which, at that time, was not incorporated and where there was religious tolerance. It was in Birmingham that the great Quaker family by the name of Cadbury set up the confectionery business that still bears their name to this day. Also Edward and Joseph Peace, who were members of a wealthy Quaker family in Darlington, were instrumental in the financing of George Stephenson in the building of the first public railway from Stockton to Darlington. It is against this background that the manufacturing industries of the UK developed in their traditional locations. Let's now look at the current situation.

Raw materials

Even with present day rapid and easy transport systems, many industries still need to be located near to sources of raw materials. Makers of iron and steel, for example, will tend to locate their factories near to the supplies of iron ore. This is because iron ore is bulky and the further it has to be transported, the higher the cost. But the industry also has to take into account the location of fuel supplies, such as coal, which is also bulky and costly to move over long distances. Almost all the iron ore used today is imported as is much of the coal used. Therefore most iron and steel manufacture is carried on mainly in coastal locations. Similarly, sugar beet refineries are located near to the beet growing areas of East Anglia and Lincolnshire, and paper making is usually sited near to conifer forests.

Human resources

The traditional industries of the UK were highly regionalized for reasons of access to raw materials, skilled labour and sources of energy such as water power and, later, coal for steam engines. The bulk transport of raw materials is no longer a problem and the electricity grid provides power wherever it is required. However, the traditional industries of cotton and woollen cloth manufacture, pottery manufacture, automobile manufacture and shipbuilding are still carried on mainly in their original locations because of the availability of a skilled and specialist workforce. The introduction of automated manufacturing processes has reduced the need for large numbers of traditionally skilled workers and manufacturing concerns are often located elsewhere for different reasons such as government financial incentives.

Incentives

When the older industries start to close down (e.g. deep mined coal) whole communities are often left without employment. To offset the deprivation that this causes, local, regional and national government together with the European Union (EU Development Fund) step in with grants and loans to set up *Enterprise Zones* to attract new industries to such areas. The incentives include low interest loans, low rents and rates, estates of ready built 'start-up' factory premises, business parks and grants. These new industries will usually be quite unrelated to the industries they are replacing and much retraining of the local labour force will be required. Government funding will also be provided for this. Therefore the traditional geographical location of industries is becoming blurred and distorted for social and political reasons unrelated to the basic logistics of manufacture.

Energy sources

The replacement of water power and steam with electricity, gas and oil coupled with modern transport systems means that manufacturing industries no longer have to be sited by rivers or coal fields. However, this does not apply to the energy providers themselves. Electricity generating stations still have to be located either close to remaining coal fields or close to special coal handling ports for cheap, imported coal. They also have to be adjacent to an adequate water supply such as a large river. Although the boiler water is recirculated, a constant supply of fresh water is required to make up the inevitable losses that occur from the cooling towers due to evaporation.

Oil terminals also have to be sited near to deep water harbours that can handle the giant oil tanker ships. Oil refineries are inevitably built adjacent to such oil terminals to ease the bulk handling of oil cargoes which can do so much damage if a spillage occurs. Natural gas is usually piped ashore and coastal sites are chosen for the purification of gas and the extraction of any useful byproducts before it is compressed and fed through the network of pipes, forming the 'gasgrid', to those areas where it is required.

Many smaller electricity power stations now use alternators driven by gas turbines similar to the jet engines of aircraft. The exhaust gases from these gas turbines, still very hot, are then used to produce steam in boilers. This steam is then used in conventional steam turbine-alternator sets to generate electricity. Such composite plants can be started up quickly to satisfy peak demands. Since they run on natural gas, they are also much cleaner and more efficient than traditional coal fired power stations. Unfortunately natural gas supplies are very limited compared with oil and coal.

Transport and markets

Transport and markets have always been interlinked. Transport has been and still is a necessary evil. It in no way enhances the value or marketability of the finished goods. If a manufacturer could have every supplier and customer next door to the factory, that manufacturer would be delighted. Unfortunately such an ideal is not possible, so transport facilities are a very important factor in locating a manufacturing plant. The need to transport raw materials and finished goods is an added cost that needs to be kept to a minimum to keep a manufacturer's products competitive.

As has been stated earlier, markets in the middle ages were essentially local events that served the surrounding community. Lack of easy and fast transport prevented perishable goods such as food being traded between centres of population which had to rely entirely upon local farmers surrounding the towns. Gradually, as the transport infrastructure improved during the industrial revolution, markets became regional and then national. Today we trade all over the world and refer to 'global markets'.

The cost of the transport of raw materials and finished goods has to be added to the cost of the product and paid for by the consumer. If the raw materials are bulkier and heavier than the finished goods, then the manufacturing plant is more likely to be sited near to the source of raw materials or the docks at which they are imported. On the other hand, if the finished goods are bulky and heavy compared with the materials and equipment from which they are made then the manufacturing plant will be sited nearer to the market–for example, it is easier to transport the steel and engines to a shipyard on the coast, than to transport the finished ship to the sea.

Transport has to be matched to the product. It is uneconomical to use air freight to transport heavy and bulky building supplies and machinery. These are better transported by road, rail or sea. On the other hand, some perishable fruit and vegetables which have to arrive fresh for the market can be economically transported by air freight. Letters and parcels can also be transported by air freight as speed is essential, as it is in the transportation of emergency medical supplies.

All manufacturing plants must be located near to appropriate transport systems. It is no use setting up a manufacturing plant just because a greenfield site has been made available in a centre of abundant cheap labour, with a government grant to build and equip the factory, unless that site is accessible for incoming raw materials and the transport of the finished products to their intended markets. Whether it is sited near a rail terminal, a road network, an airport or a seaport (docks) or a combination of these will depend upon the type of raw materials used and the type of products manufactured. The choice of transport will ultimately depend upon such factors as:

- weight;
- bulk;

- fragility;
- perishability;
- quantity;
- urgency;
- cost and value.

Therefore communications and ease of access to the market are vital considerations. A company which imports its raw materials from Scandinavia or which exports its products to Scandinavia, for example, will locate its factories in areas with good communications to ports on the east coast of England and Scotland. Companies trading with Europe will tend to site their factories in the south east of the UK to take advantage of the cross Channel ferries. However, with the advent of the Channel Tunnel and the opening of rail freight terminals in the industrial Midlands and elsewhere in the UK and linked with the Channel Tunnel, this location of exporting businesses adjacent to ports is no longer so vital.

This first element of Chapter 1 only covers a few of the larger manufacturing industries in the UK. In addition there is the ceramics industry making goods ranging from bricks, roofing tiles and drainpipes to ceramic tiles, sanitaryware and tableware. There are also the wood working industries producing goods ranging from joinery items like window frames, door frames, doors and staircases to the finest furniture making. No book could cover every branch of manufacturing still carried on in the UK.

Test your knowledge 1.9

1. List the main factors that influence the siting of a manufacturing company.

2. Name the four major iron and steel making plants in the UK, state where they are located and discuss the reasons that dominated their choice of location.

3. Suggest reasons why a major tobacco products firm is located at Bristol.

4. Suggest reasons for the siting of the Imperial Chemical Industries chemical plant at Billingham.

Key Notes 1.1

- Manufacturing is the conversion of raw materials into useful articles by means of physical labour or the use of power driven machinery.
- The purpose of manufacturing goods is to satisfy demand and create wealth.
- Currently some 4.3 million people, representing 20% of all employed persons in the UK, are employed in the manufacturing industries.
- Manufacturing accounts for 80% of all exports and 20% of the Gross Domestic Product (GDP)–see below.
- Gross Domestic Product (GDP) is the total value of all the goods manufactured in the UK no matter who are the owners of the factories or where in the world they live or are based.
- Gross National Product (GNP) is the total value of all the products manufactured by companies owned by the citizens and financial institutions of the UK no matter where in the world such plants are sited.
- Balance of trade is the difference in value between the UK imports and exports for a given period of time, usually monthly or yearly.

- The main manufacturing sectors in the UK are the: chemical industry; engineering industry; food, drink and tobacco industry; paper and board manufacturing industry; printing and publishing industry; textile, clothing and footwear industry.
- The engineering industry may be broken down into the following sectors: electrical and electronic engineering; instrument engineering; mechanical engineering; transportation equipment engineering.
- The mechanical engineering sector may, itself, be broken down into the following subsectors: agricultural equipment engineering; component manufacturing; construction equipment manufacturing; machine tool manufacturing; mechanical handling equipment manufacturing; miscellaneous machinery and accessory manufacturing; office machinery manufacturing; plant manufacturing.
- The main factors influencing the location of industries are: a suitable and available site; available raw materials; a plentiful supply of labour with the appropriate skills; adequate and low cost transport; easy access to markets; financial and political inducements (government loans and grants); adequate and low cost energy sources.

Evidence indicator activity 1.1

1. Write a report that describes the key features of each of the six main manufacturing sectors. For each sector:

   - include at least two examples of their products;
   - identify their regional locations;
   - compare the relative economic importance of the sectors.

2. Write a report that considers, in general terms, how resources, incentives and infrastructure influence decisions about where companies locate their operations. The report should be supported by two examples of actual companies which decided to locate in particular places but for different reasons.

1.2 Production systems

1.2.1 Key stages of production

As we saw in the previous element, the finished product of one industry is often the raw material of another industry. Also, that the waste product of one industry can be the essential raw material of yet another industry. In this way many quite different manufacturing industries can be closely linked as suppliers and users of each other's products and of each other's waste. Therefore it is often difficult to decide exactly where a *production system* starts and finishes.

For the purposes of this book we will try and simplify the *key stages of production* by limiting ourselves to the processes that occur *within a single plant*. For example, let's consider a loaf of bread. For simplicity we will ignore the growing of the corn and the milling of the grain to make the flour. Let's only consider what happens in the bakery and see how the *key stages of production* shown in Fig. 1.24 can be applied to our loaf of bread. Remember that the grain from the farmer is the raw material for the flour miller and the flour from the mill is one of the raw materials for the baker. Also, byproducts from flour milling are the natural gums dextrin and gluten. These are the raw materials for the adhesives used in the manufacture of envelopes, stamps and food labels. Being safe natural products there is no possibility of toxic contamination in their use.

Note: 1. Not all products require all these stages.
2. Not all products require all the stages in this order.

Figure 1.24   Key stages of production.

Material preparation
The various ingredients for making our loaf will be checked for quality and freedom from contaminants. The ingredients will be measured out into the quantities required for the type of bread being made. Typically 250 kilograms of flour, 140 litres of water, 4 kilograms of salt and 2.5 kilograms of compressed yeast to make a batch of the basic dough.

Processing
The above ingredients are kneaded into a smooth dough by machinery and various other ingredients such as fat, milk powder and sugar are added according to the type of bread being made. The dough is then allowed to stand and ferment for several hours, after which it is automatically divided up and machine moulded to shape before it is baked for approximately three-quarters of an hour at a temperature of 230°C. The bread is then allowed to cool.

Finishing
In this example we can assume that we require a sliced loaf. Therefore the finishing process will be that of automatically slicing the loaf by machine.

Packaging
Since we are considering the manufacture of a sliced loaf, the bread will be automatically wrapped to keep it fresh. The packaging will be printed with the manufacturer's logo for advertising purposes together with the list of ingredients and nutritional information now required by law. A flow chart for key stages in the manufacture of our loaf of bread is shown in Fig. 1.25.

Let's now consider the manufacture of the electrical conduit fitting shown in Fig. 1.26(a). It consists of a malleable cast iron body with a cover plate stamped out of sheet steel, retained in position by two screws. The key production stages are shown in Fig. 1.26(b). You will see that there is some variation in the sequence of the production processes. This is because there must be no paint in the screw threads from the finishing process. In this example, such painted threads would make it difficult to assemble the product. The paint would also act as an electrical insulator and prevent metal to metal contact between the metal conduit and the metal fittings. This would

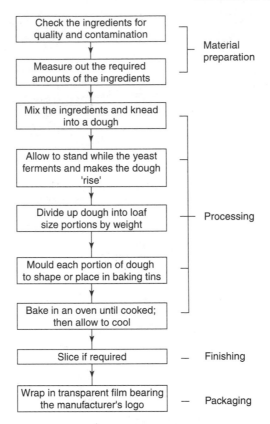

Figure 1.25 Key stages in manufacture of a loaf of bread.

be very dangerous because the lack of metal to metal contact would destroy the continuity of the earth return path essential for electrical safety.

Material Preparation

The body of the fitting is manufactured from malleable cast iron and will have its surface covered in scale and some loose sand from the mould. It will also have a blemish where the 'pouring gate' was cut off.

- First, any surface blemishes left over from the casting process will need to be ground off.
- Next, any residual sand and scale will need to be removed by grit blasting or by tumbling in large steel barrels along with a suitable grit.

The steel for the cover plate will be supplied in strip form in coils and will need to be loaded onto a suitable stand for feeding into the stamping press.

The screws will be 'bought in' ready made but will need to be checked that they are the correct length and have the correct screw thread.

Processing (cover plate)

The cover plate will be stamped from the steel strip in a *progression* type press tool. As the metal 'progresses' through the tool the screw holes are stamped

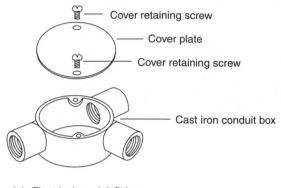

(a) Electrical conduit fitting

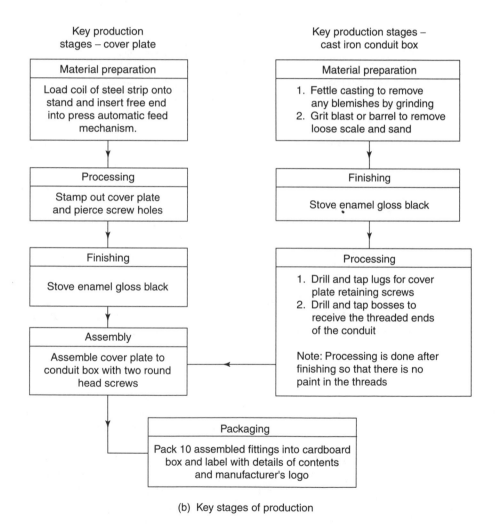

(b) Key stages of production

Figure 1.26 Manufacture of conduit fitting.

through the metal in a 'piercing' operation. Then the cover plate—complete with holes—is stamped out of the strip in a 'blanking' operation. This is a repetitive operation and the stamping press does not stop until the end of the coil of steel strip is finally reached.

Finishing

For this particular product, finishing takes place before final processing and assembly. Both the castings and the cover plates will be finished gloss black by stove enamelling.

Processing the castings

The castings will be drilled and *tapped* to take the retaining screw for the cover plates. Note that 'tapping' is the name given to cutting internal screw threads. The bosses will also be threaded by 'tapping' so that the metal conduit tube can be screwed into them.

Assembly

The cover plates will be assembled to the body castings by suitable screws. If these are brass, they will not have undergone a finishing process but will be 'self-colour'. If they are steel they will have black-enamelled heads. Remember that the key stages of production will always be present but the order in which they appear in the production sequence will vary from product to product.

Let's now broaden our approach and see how the *key stages of production* apply to other manufactured goods.

Material preparation

The inspection and preparation of the incoming materials ready for processing are very important in achieving and maintaining the quality of the finished product. Quality control will be dealt with in Section 1.2.4. Some further examples of material preparation are:

- The washing and grading of fruit and vegetables in the food processing industries.
- The unpacking and thawing out of frozen produce such as fish and poultry before processing.
- The cutting of cloth to size and shape in the clothing industry and the cutting of leather to size and shape in the footwear industry.
- The blending, cutting and shredding of leaf tobacco ready for the manufacture of cigarettes and cigars.
- The degreasing and surface treatment of metal products before processing.

Processing

Manufacturing processes are used to convert materials from one form to another. As we have already seen, a bakery converts flour, water and yeast into bread. Some further examples of processing are:

- The casting of metal ingots into useful shapes such as automobile cylinder blocks in a foundry.
- The cutting of components from steel bars in the machine tools of an engineering workshop.
- The forming of plastic moulding powders into finished products by compression moulding and injection moulding machines in the plastics industry.
- The slicing and precooking of potatoes to make 'oven-ready' chips in the food industry.
- The machining of wood to make the component parts of doors, window frames and staircases in the joinery industry.

Finishing

Various finishes are applied to manufactured goods to give them either an attractive appearance or to protect them from deterioration or both. Some examples are:

- The painting of exterior woodwork. This improves the appearance and also preserves the wood and protects it from the onset of wet rot. The paint system should consist of a primer, an undercoat and a topcoat (gloss) containing a varnish to seal the system against the penetration of moisture.
- The painting of steelwork to prevent rusting. This requires a different type of primer paint containing corrosion inhibitors rather than the fungicide of a wood primer. In both cases the primer provides a 'key' to help the undercoats and topcoat adhere to the object being painted.
- Galvanizing. Goods manufactured from steel are dipped into molten zinc which forms a corrosion resistant coating on the steel.
- Electroplating such as nickel plating and chromium plating. This is less corrosion resistant than galvanizing but has a more decorative appearance.
- Anodizing aluminium where an aluminium oxide layer is built up on the metal. Sometimes the chemicals used provide an oxide film of the required colour. Sometimes the film is coloured after anodizing.

As well as providing protection and improving the appearance of manufactured products, finishing processes can be used for other purposes. For example:

- Waterproofing fabrics for manufacturing wet-weather clothing.
- Liquid repellent treatments for carpets to protect them against staining when liquids are accidentally spilt onto them.
- Fire retardant and flame resistant treatments for furnishing and clothing fabrics.
- Crease resistant and drip-dry treatments for clothing fabrics. Essential for the clothes of people who travel a lot.

Assembly

This is the joining together of various manufacture parts to make a complete product. Sometimes the assembly is permanent as when the body panels of a car are welded together. At other times the assembly is temporary as in the bolted-on wheels of a car. It must be possible to quickly and easily remove and change a wheel if a tyre becomes punctured. Some examples of assembly processes are:

- The riveting and welding together of engineering structures to make bridges, ships and offshore oil rigs.
- The use of synthetic adhesives in the manufacture of goods ranging from wooden furniture to the stressed components of aircraft.
- The stitching together of clothing.
- The soldering of electronic components to the printed circuit boards in products such as computers and television sets.

Packaging

The packaging of finished products has become an important part of the marketing process. The requirements of packaging are:

- To protect the goods from damage in transit.
- To protect the goods from contamination.
- To protect the goods from climatic changes.
- To enable the goods to be stored correctly and without damage.
- To enable the goods contained within the packaging to be easily identified without the need for unpacking.
- To be informative and attractive to the customer.
- To comply with legislation.

Very often the packaging is the first line of communication with a prospective customer. Therefore the packaging is an important means of communicating and selling. It must be pleasing in design and it will always contain the brand name. The busy shopper in a supermarket must be able to locate and identify a particular product and brand instantly on the crowded shelves. To comply with legislation it must also state the country of origin. Food products especially are subject to legislative requirements such as nutritional information and the ingredients used.

Typical packaging materials consist of paper and boards, and rigid, flexible and foam plastics. Foam plastics are frequently moulded to the shape of the product so that it is fully supported. Liquids may be contained in moulded plastic or glass bottles or in metal cans. Solid foods may be contained within metal cans (for example, corned beef) or in plastic prepacked trays covered with a transparent film (for example, bacon).

Finally, as we have seen in this section, not all the key stages are used every time. You do not have to assemble a loaf of bread. Nor do the stages always follow in the same order. The various parts of a car are painted and/or electroplated (finished) after subassembly but before final assembly.

Test your knowledge 1.10

1. List the key stages of production.

2. Describe briefly how these key stages may be applied to a product of your own choice.

1.2.2 Scales of production

Popular makes of motor cars are produced in large quantities by increasingly automated manufacturing techniques because of the large demand for such products. On the other hand, large structures such as bridges are usually built on a 'one-off' basis mainly by hand. However, some components, such as nuts and bolts, used in the construction of a bridge may well be mass produced on automatic machines because of the quantity required. Top restaurants will generally employ chefs to cook meals to order to a very high standard for a limited number of customers who are able and willing to pay the high cost of such service. On the other hand, fast food outlets are more likely to use processed foods that require the minimum of preparation. This increases the volume of meals that can be served and reduces the time and cost involved.

Therefore the *scale of production* will depend upon the type of product and the demand for that product. The broad groupings are as follows:

- Continuous flow and line production.
- Repetitive batch production.
- Small batch, jobbing and prototype production.

Let's now look at some typical applications of each category.

Continuous flow and line production

In *continuous flow production* the plant resembles one huge machine in which materials are taken in at one end and the finished products are continuously despatched from the other end. The plant runs for 24 hours a day and never stops. Plastic, glass sheet and plasterboard are produced in such a manner.

In *line* or *mass production* plants the manufacturing system plant is laid out to produce a single product (and limited variations on that product) with the minimum of handling and where the product is moved from one operation and/or assembly station to the next in a continuous, predetermined sequence by means of a conveyor system. Individual operations are frequently automated. Such plants usually manufacture consumer goods such as cars and household appliances in large quantities in anticipation of orders. The layout of a typical flow or line production plant is shown in Fig. 1.27.

The characteristics of *continuous flow* and *line production* can be summarized as:

- High capital investment costs.
- Long production runs on the same or similar product type. In some cases production never ceases until the plant requires refurbishment or the product changes.
- Highly specialized plant resulting in inflexibility and difficulty in accepting changes in product specifications.
- Rigid product design and manufacturing specifications.
- The processing equipment is laid out to suit the operation sequence.
- Stoppages in any one production unit need to be rectified immediately or the whole production line is brought quickly to a halt.

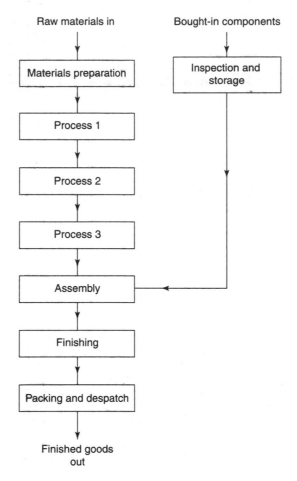

Figure 1.27 Layout of flow production plant.

- Components, products and subassemblies move from one workstation to the next by means of pipelines in the case of fluids, gases and powders or by mechanical conveyors in the case of solid goods.

Typical products requiring this type of production are shown in Fig. 1.28.

Batch production

As its name implies, this involves the production of batches of the same or similar products in quantities ranging from, say, 10 units to several hundred units. These may be to specific order or in anticipation of future orders. If the batches of components are repeated from time to time, this method of manufacture is called *repetitive batch production*. General purpose rather than special purpose machines are used and these are usually grouped according to process as shown in Fig. 1.29.

Nowadays, the machines are often arranged to form *flexible manufacturing cells* in which the machines may be computer numerically controlled (CNC) and linked with a robot to load and unload them as shown in Fig. 1.30.

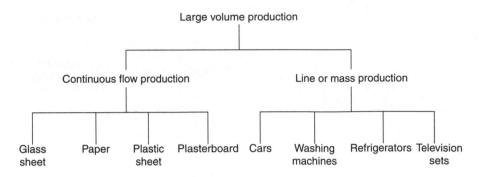

Figure 1.28 Large volume production.

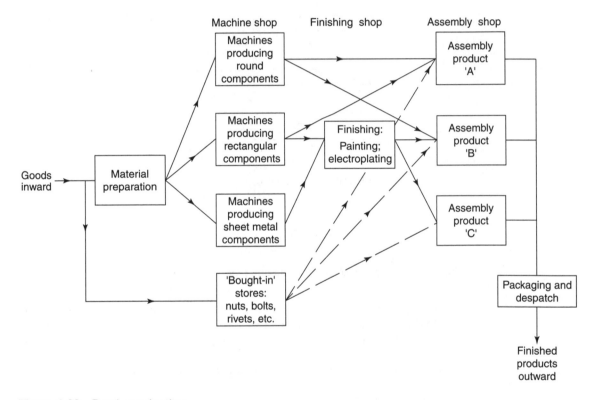

Figure 1.29 Batch production.

To change the product, the computers are reprogrammed. The computer programs are kept on disks and are available for repetitive batches thus saving lead time in setting up the cell.

The characteristics of batch production can be summarized as follows:

- Flexibility. A wide range of similar products can be produced with the same plant.
- Batch sizes vary widely.

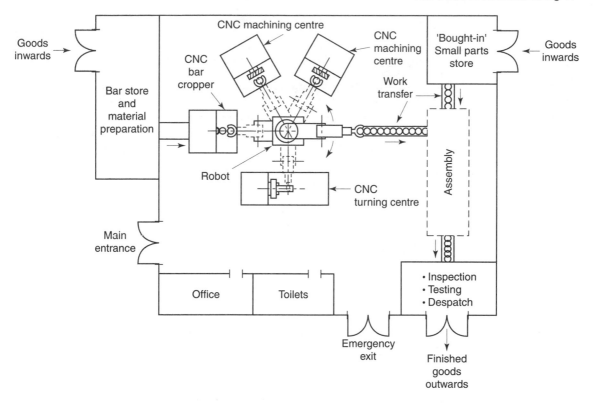

Figure 1.30 Layout for large batch production.

- Production is relatively slow as one set of operations is usually completed before the next is commenced.
- Work in progress has to be stored between operations. This ties up space and working capital. It also has to be transported from one group of machines to the next.
- General purpose machines are used. These have a lower productivity than the special purpose machines used in line and flow production.
- It is unlikely that all the machines in any particular plant are required for every product batch. Therefore work loading will tend to be intermittent with some machines remaining idle from time to time, while others are overloaded resulting in production bottlenecks.
- For the reasons stated above, unit production costs are very much higher than for flow and line production. Process efficiency is lower since variations will exist between the rate of production and consumer demand.
- Batch production is the most common method of manufacture in the UK.

Some typical product groups are shown in Fig. 1.31.

Small batch, prototype and jobbing production

This refers to the manufacture of products in small quantities or even single items. The techniques involved will depend upon the size and type of the

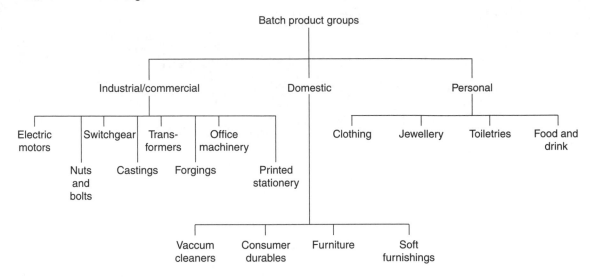

Figure 1.31 Some typical batch product groups.

product. For very large products such as ship, oil rigs, bridges and the steel frames of large modern buildings, the workers and equipment are brought to the job. At the other end of the scale, a small drilling jig built in a toolroom will have parts manufactured on the various, specialist machining sections and will be brought to and assembled by a specialist toolroom bench-hand (fitter). Prototypes for new products are made prior to bulk manufacture to test the design specification and ensure that they function correctly. Modifications frequently have to be made to the prototype before manufacture commences.

Workshops for small batch (10 or less) and single products such as jigs, fixtures, press tools and prototypes are referred to as *jobbing shops*. That is, they exist to manufacture specific 'jobs' to order and do not manufacture on a speculative basis. The layout of a typical jobbing shop is shown in Fig. 1.32.

So far we have only considered an engineering example, but the same arguments apply elsewhere. For example, when you order a suit from a bespoke tailor it will be manufactured as a 'one-off' specifically to your measurements and requirements. It will be unique and made mainly by hand in the tailor's workroom. The tailor will not manufacture on a speculative basis. However, the suit you buy from a clothing store will be one of a *batch* produced in a factory in a range of standard sizes and a range of standard styles. The various parts of the suit will be cut out and made up by specialist machinists.

The characteristics of small batch and prototype production can be summarized as:

- Work is quoted for job by job. Nothing is manufactured for stock.
- A wide range of general purpose machines and associated processing equipment (e.g. heat treatment equipment for a toolroom) is required.

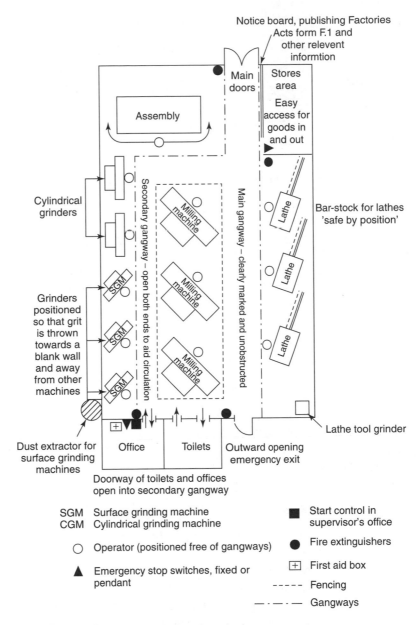

Figure 1.32 Jobbing shop layout (engineering).

- Highly skilled and versatile operators are required.
- Economical loading of the plant and personnel is difficult.
- Often, only outline specifications are provided, therefore design facilities are required.

Some typical products produced on a small batch or a jobbing basis are as shown in Fig. 1.33.

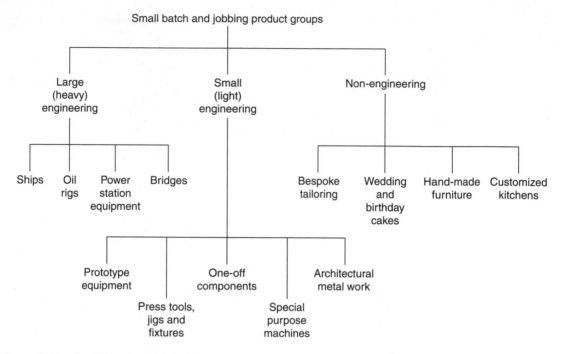

Figure 1.33 Small batch and jobbing product groups.

Test your knowledge 1.11

1. Briefly summarize the essential differences between:

   (a) continuous (flow) production;
   (b) batch production;
   (c) jobbing production.

2. Give an example where each of the production methods listed above would be appropriate and state the reasons for your choice.

1.2.3 Production systems

All production systems have a single aim, to convert raw materials into a finished product that can be sold at a profit. The *inputs* and *outputs* of a typical manufacturing system are shown in Fig. 1.34. These can vary in detail depending on the product being manufactured. Note how quality control covers the whole system.

Let's now consider the *inputs* in greater detail.

Sourcing and procurement

Sourcing is finding suitable sources of raw materials for a particular manufacturing process. It can range from scanning trade directories and the *Yellow Pages* to global exploration for mineral resources such as coal, crude oil and metal ores.

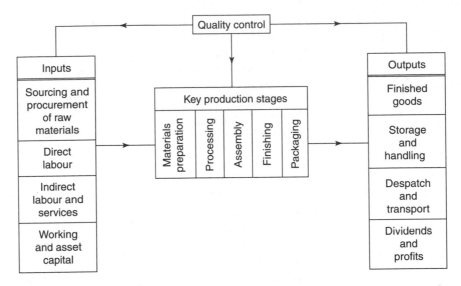

Figure 1.34 A typical production system.

Procurement is the process of acquiring the raw materials once they have been sourced. This involves negotiating a suitable price and scheduling the delivery quantities and dates. The supplier usually arranges suitable transport.

Labour
The labour, both skilled and unskilled, associated with the manufacture of any product can be broken down into two categories.

- *Direct labour*. This refers to the operatives and other personnel directly employed on a particular job and whose time can be charged directly to that job.
- *Indirect labour*. This refers to the administrative and sales staff needed to run the company and whose time cannot be directly apportioned to a particular job.

Nevertheless the selling price of the job must incorporate a contribution towards the indirect labour costs.

Services
These refer to the supply of oil, gas, water, electricity, telephones, facsimile transmissions, electronic mail and mail delivery and collection, for example. All these have to be paid for. Again, it is difficult to assign these costs to particular jobs. Nevertheless the selling price of the job must incorporate a contribution to the service costs.

Capital
This is the money raised to run the company. It can come from various sources depending upon the size of the company. For instance:

- From the savings of the founder of the company in the case of new small companies.
- From bank loans and finance houses.
- From the sale of shares on the stock market in the case of public limited companies.
- From retained profits from the sales of goods.

Note the more a company borrows the more it has to pay out in interest charges and dividends.

Capital can be divided into two categories:

- *Asset capital* representing the money invested in the premises and plant. Allowance must be made for the gradual deterioration and obsolescence of the premises and plant and money must be set aside for its eventual replacement.
- *Working (trading) capital*. This is the money required to buy the raw materials and pay for the wages and services until the goods are completed, delivered and paid for. If this money has to be borrowed, the interest that has to be paid can easily wipe out the profit earned by a particular job if the customer is slow in paying.

Assets

These have already been introduced and represent the premises, plant, machinery, tools, vehicles, disposable raw materials and finished products, cash on deposit at the bank and 'goodwill'. Items that should not be considered as assets include Value Added Tax (VAT), Pay As You Earn (PAYE) income tax and National Insurance (NI) money collected but not yet paid over to the government bodies concerned. These funds do not belong to the company which only acts as a collecting agency. Similarly, pension funds should not be counted as a capital asset since they are held in trust for the persons contributing to such funds.

The *key production* stages have already been considered in Section 1.2.1, so let's move on to the *outputs* or, as they are sometimes called, *outcomes*.

Finished products

These are the manufactured products that are produced from the raw materials *procured* by the purchasing department, processed through the *key production stages* and prepared ready for *despatch* to the customer.

Despatching

This is involves the following procedures:

- Packaging the finished products to prevent damage in transit.
- Goods going abroad may need to be tropicalized if they pass through or are destined for areas of high temperatures and high humidity.
- Preparing any documentation that must accompany the finished product. This is particularly important in the case of export orders and also in

the case of hazardous products (e.g. chemicals that may be flammable and/or toxic).

- Arranging transport appropriate for the type of product and the journey it has to take. In the case of large products it may be necessary to arrange a police escort and route the transport so as to avoid low bridges.
- Advising the customer of any special offloading facilities that may be required. For instance site accessibility and/or the need to hire a mobile crane.
- Finally, loading the finished product at the works and sending off to its final destination.

Storage

Where products are made speculatively instead of to special order, the goods will need to be stored while waiting despatch to the customer. Such storage (warehousing) facilities should be appropriate for the type of product concerned. Overhead cranes may be required where machine tools are being stored and handled. Room may be required to manoeuvre fork-lift trucks if consumer durables such as washing machines, cookers and refrigerators are being warehoused. Stationery and office supplies may be dealt with manually. Air conditioning and ventilation may be appropriate to avoid deterioration of the stored goods. Finally, the stores should be organized on a 'first-in, first-out' basis to ensure a constant circulation of goods and to avoid overlong periods of storage during which deterioration can set in.

Profit

Manufacturing is carried out for two purposes:

- *To satisfy a demand*. That is, to produce the goods that the customer wants.
- *To make a profit* for the owner or owners of the plant. The owner may be a single individual, a family or, in the case of a limited company, the shareholders.

The profit is the money left over from the sale of the finished product after all the expenses involved in its manufacture have been paid. This is shown in Fig. 1.35.

Quality control

At one time the only quality control consisted of a final inspection before despatch. Nowadays quality control is a continuous process involving the suppliers of raw materials as well as the manufacturer. This enables any failure due to faulty materials or manufacture to be traced to its source. It is now a total company responsibility with every worker involved. This is why quality control has been shown to embrace all the activities shown in Fig. 1.34. Quality control will be considered in detail in Section 1.2.5 of this chapter.

Finally Fig. 1.36 shows a wooden stool and Fig. 1.37 shows how a production system flow diagram is applied to its manufacture.

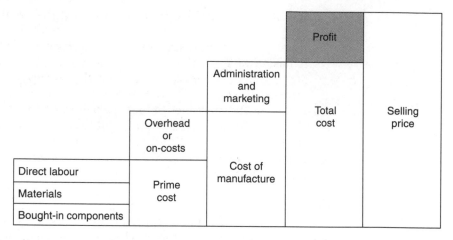

Figure 1.35 Cost structure.

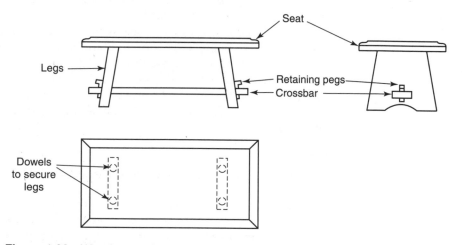

Figure 1.36 Wooden stool.

Test your knowledge 1.12

1. Briefly explain what is meant by a production (manufacturing) system and list its key elements.

2. State the essential differences between 'sourcing' and 'procurement'.

3. Explain the difference between 'direct labour' and 'indirect labour'.

4. State the essential difference between 'asset capital' and 'working (trading) capital'.

5. Explain what is meant by 'profit' and suggest reasons why profits are essential for the well-being of a company.

1.2.4 Changing scales of production

Changing the scale of production has far reaching effects on the method of manufacture employed and on the cost of the products made–for example, a component for a car that is mass produced can be made relatively cheaply.

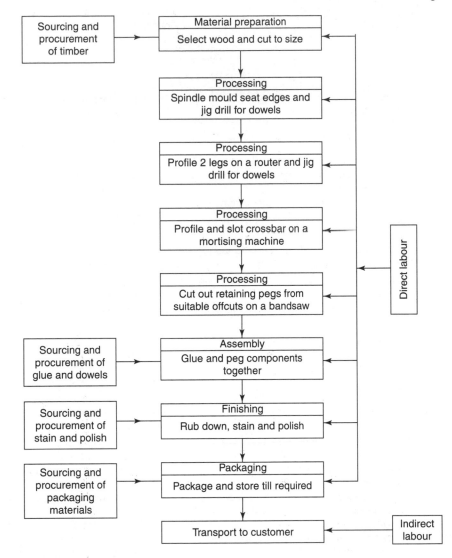

Figure 1.37 Production system for wooden stool.

However, the years pass by and, eventually, that car and its component parts will no longer be in production. Some day it may become a collector's item and will be restored by an enthusiast. Spare parts will now have to be made on a one-off basis. Compared with the original cost of the mass produced parts, these one-off replacements will be very much more expensive and no longer available off-the-shelf. The effects of changing the scale of production are shown in Figs 1.38, 1.39 and 1.40.

Test your knowledge 1.13

1. Explain briefly what is meant by the scale of production.

2. State the main effects of changing the scale of production for a manufactured article of your own choice.

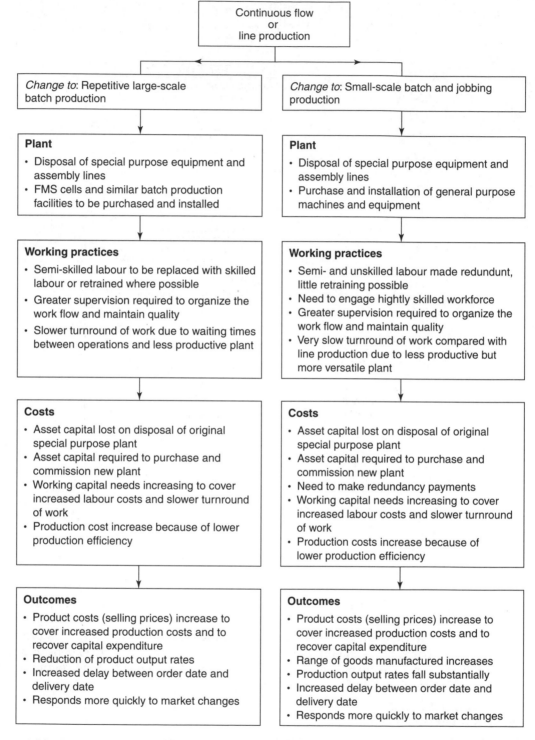

Figure 1.38 Effects of changing from flow and line production.

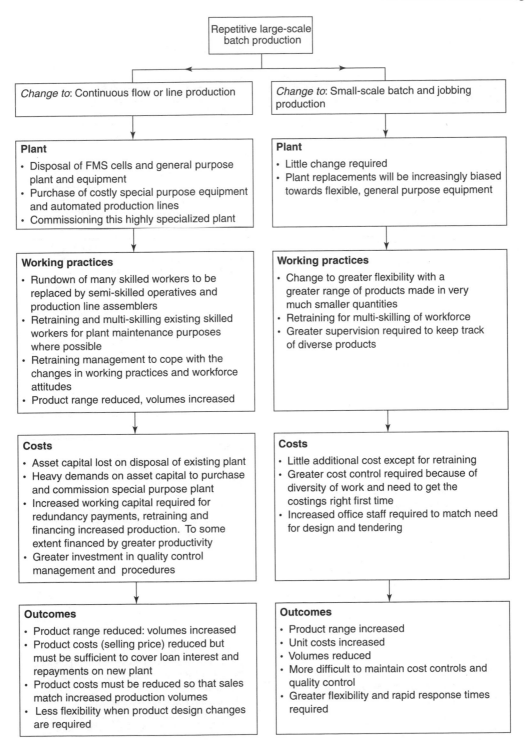

Figure 1.39 Effects of changing from repetitive large-scale batch production.

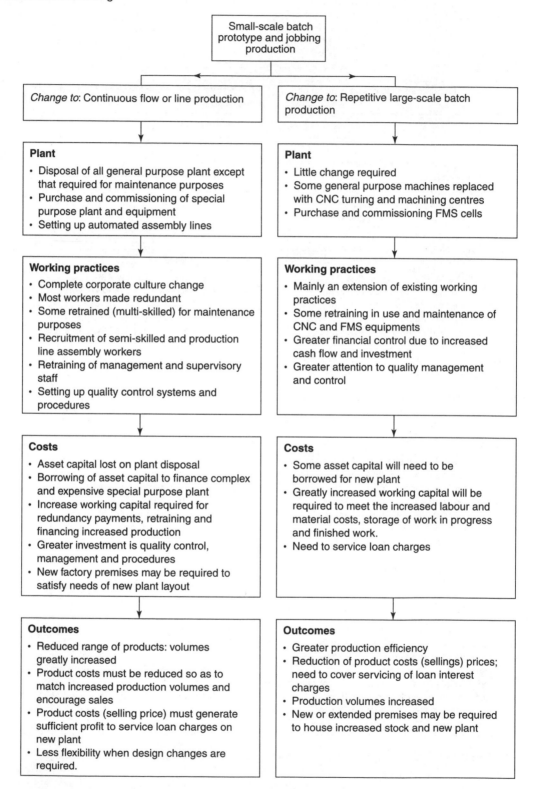

Figure 1.40 Effects of changing from small-scale and jobbing production.

1.2.5 Quality control

Quality can be defined as *'fitness for purpose'*. There is no such thing as absolute quality. A customer's ideas of what represents quality will change with time and with competition in choice. A motor cycle that was considered state of the art 50 years ago would not satisfy the requirements of the present day market except, perhaps, as a collector's item.

Therefore any manufacturing company's attitude to quality must be constantly reviewed to satisfy changes in the expectations of its customers. At all times a company's products must represent:

- Value for money.
- State of the art technology.
- Fitness for purpose.

Quality assurance

Quality assurance is the result of creating and maintaining a quality management system that insures all finished products achieve *fitness for purpose* and *conform* to the specifications agreed with the customer.

In the UK, ISO 9000 lays down the guidelines and rules for a quality management system that is recognized and approved internationally. Since companies that are ISO 9000 approved can only purchase their goods and services from similarly approved companies, it is necessary for most companies to be approved. To obtain the ISO 9000 kite mark a company must have its quality management system assessed by an independent, external certification organization. If it passes this *audit* and conforms to the requirements of the national standard, its name appears in the Department of Trade and Industry (DTI) *Register of Quality Assessed United Kingdom Companies*. It can then trade with other similarly accredited companies and its customers can be *assured* of the quality of its products.

The benefits of satisfying the requirements of ISO 9000 and being registered with the DTI can be summarized as:

- Greater control over raw materials and bought-in components.
- A complete record of production at every stage to assist in product or process improvement.
- Cost effectiveness because there is a reduction in scrap and waste and the need for reworking. Products are increasingly *right first time*.
- Customer satisfaction because quality has been built in and monitored at every stage of manufacture before delivery. This enhances a company's reputation and leads to repeat orders and increased profitability.
- Acceptability of a company's products in global markets.

Let's now see how this can be achieved.

Quality control

Quality control is achieved by setting up a system, under the direction of a *quality manager*, that embraces all members of a company from the chairman of the board downwards and has the full support and commitment of the board and senior management. The quality system must provide for:

- *Traceability* so that all products, processes and services needed to fulfill a customer's requirements can be identified in order that they can be traced throughout the company. This is necessary in the case of a dispute with a customer regarding the quality of any particular product supplied or component within that product.
- *Control of design* so that any design meets a customer's requirements through consultation by the design and marketing staff with the customer.
- *Control of 'bought-in' parts* to ensure that they conform to previously agreed specifications. This verification can be achieved by buying only from an ISO 9000 accredited source or by rigorous inspection of the bought-in parts.
- Control of manufacture through clear work instructions and documentation together with effective process control and inspection procedures. Accreditation of the company's measuring and testing equipment should be by an external body such as UKAS (United Kingdom Accreditation Service).

Let's now see how these aims involving measuring, recording and maintaining the required standards of production can be achieved.

It would be too expensive to inspect every component at every stage of its manufacture. Therefore when batch, flow and line production is undertaken it is necessary to use statistical sampling techniques, to establish *points of inspection* and to specify *quality indicators*. Fig. 1.41 shows a typical production system and the position of the *points of inspection*. These are also called *quality control* points.

Points of inspection

These must be inserted into the manufacturing chain so that costly time is not wasted on processing parts that are already defective. There is no point in gold plating a defective watch case. Such inspection points should be:

- Prior to a costly operation.
- Prior to a component entering a series of operations where it would be difficult to inspect between stages (e.g. within an automated flexible manufacturing cell).
- Prior to a processing station where failure of a defective part could cause an enforced shutdown of the whole station (e.g. automatic bottling plants, where breakage of a bottle could result in machine damage or failure, in addition to the time lost and cost of cleaning the workstation).
- Prior to a *'point of no return'* were rectification would be impossible after the operation (e.g. final assembly of a *'sealed-for-life'* device).

Points of inspection will vary from plant to plant and from product to product but there are certain points of inspection that have reasonably common ground.

Raw materials and bought-in parts. As previously stated these should be bought preferably from an ISO 9000 accredited source and be manufactured to an established BS/ISO specification. If not, then *acceptance sampling and testing* should be carried out on every batch procured. Some key items and

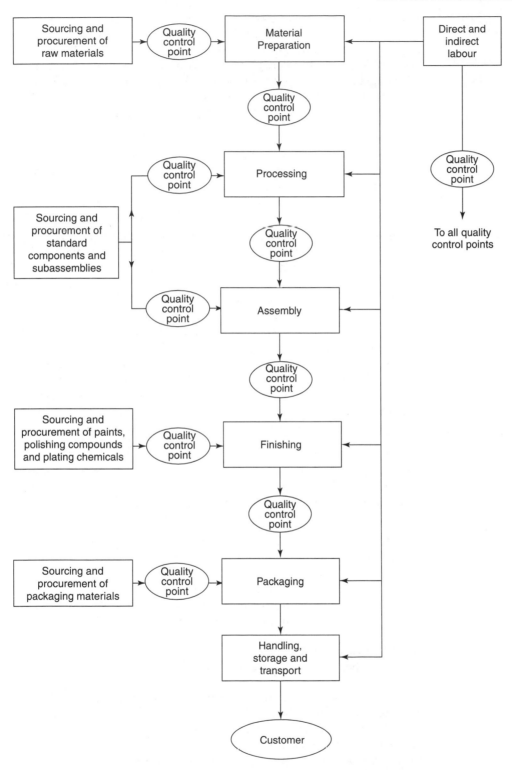

Figure 1.41 Typical production system showing the quality control points.

materials, such as those used in vehicle and aircraft control systems, will undergo acceptance sampling irrespective of their source of procurement. Suppliers are usually given a *vendor rating* depending on the percentage of acceptable products received previously and the number of warranty claims made against the supplier. On no account must substandard materials and parts be allowed into the production chain as this may lead to the whole batch being scrapped after expensive processing has been carried out.

- *Material preparation*. This is the next point of inspection to ensure that the correct quantity of the correct material has been issued for the job. Also that it has been cut to the correct size and any degreasing or other preliminary preparation process has been carried out.
- *Processing*. Modern manufacturing organizations usually aim to produce goods that are *right first time*. You cannot inspect quality into a product. Inspection can only hope to remove defective (faulty or flawed) components from the manufacturing sequence. These defective products then have to be corrected or scrapped. This costs the company money. Therefore it is more economical to ensure than products are made correctly from the start, that is, they are made *right first time*.
- *Assembly*. Even assuming that all the components parts are correctly manufactured, it is still essential that they are correctly assembled. It is important that components have not been omitted or damaged during assembly, or that electric wiring has not been incorrectly connected.
- *Finished product*. The final inspection must include correct appearance (freedom from blemishes), correct functioning of the assembly, total compliance with the customer's specification and total compliance with national and international safety and legal requirements.

The quality control department, under the leadership of the quality manager, and with the full support of the board of directors, needs to establish well-defined procedures at each point of inspection in order to avoid confusion and to ensure that the system works consistently. These procedures should include:

- The method of inspection and the sampling frequency.
- The reworking, downgrading or scrapping of components and assemblies.
- The recording and analysis of inspection data.
- The feedback of information so that faulty manufacturing and assembly processes and techniques can be corrected in the aim to be *right first time*.

Quality indicators

Quality indicators can be categorized as:

Variables. These are characteristics where a specific value can be measured and recorded and which can *vary within prescribed limits*. For example, a particular component dimension is given as 100 ± 0.5 mm. Therefore any component with a measured dimension lying between 100.5 mm and 99.5 mm inclusive would be acceptable. Variables are measurable and can include, for example:

- length, width, height, diameter, position and angles inspected by measurement;
- mass;
- electrical potential (volts), current (amps) and resistance (ohms);
- fluid pressure;
- temperature.

Attributes. These can only be acceptable or unacceptable – for example, colour (if you have ordered a blue suit, a brown one will not be acceptable).

Quality indicators will be discussed more fully in Chapter 3. However, we can list them briefly as including:

- length, width, height, diameter, radius, area and volume;
- appearance;
- taste, sound, smell and touch;
- functionality (fitness for purpose).

Inspection and testing methods

Inspection and testing are essential to all quality control systems. However, the inspection and testing techniques adopted will depend upon the product. The inspection of a shirt or a pair of trousers will be far less rigorous than the inspection of an aircraft engine. Further, the inspection process used will depend upon the position of the quality control point in the manufacturing system.

Materials inspection can include:

- mechanical testing for strength;
- chemical analysis;
- sonic and X-ray testing for cracks and flaws;
- testing for physical properties (electrical, magnetic and thermal);
- visual appearance.

Process inspection can include:

- dimensional measurement and gauging;
- surface finish measurement;
- visual appearance;
- balance of rotating parts.

Final inspection can include:

- visual appearance;
- function under normal working conditions;
- conformance testing.

Data recording

As has been stated previously an important function of ISO 9000 is traceability so that, in the event of a malfunction, the history of a faulty component or assembly can be traced back to its source. Therefore record keeping is of vital importance in a quality control system. Inspection data recording will be dealt with in greater detail in Chapter 4 but can be summarized here as:

- Hand written records on standard report forms. Often, these merely require a series of prompts to be ticked off on a *satisfactory/not satisfactory* basis.
- Computer entries onto a database.
- Computer compilation of tables and/or graphs of statistical data from measured variables. These are extremely valuable in process control and are essential in achieving *right first time* manufacture.

Test your knowledge 1.14

1. (a) Explain what is meant by 'quality'.

   (b) Explain briefly why quality control is so important in modern manufacturing industries.

2. Explain what is meant by 'quality assurance'.

3. Explain what is meant by the term 'points of inspection' and indicate where they should be introduced into a production (manufacturing) system.

4. Explain briefly why it is important for manufacturing companies to be ISO 9000 accredited.

5. In your own words, explain the essential differences between 'variables' and 'attributes' as quality indicators.

6. Explain briefly why record keeping is essential to a quality control system.

Key Notes 1.2

- The key stages of production are: material preparation; processing; assembly; finishing; packaging.
- The scale of production can be: continuous flow; line (mass) production; batch production; small batch and jobbing production.
- Sourcing is finding suitable sources of raw materials.
- Procurement is negotiating the purchase and delivery of raw materials.
- Direct labour refers to those persons directly employed in the manufacture of an article and whose time can be directly charged to that article.
- Indirect labour refers to persons whose time cannot be charged directly to the manufacture a particular article, for example: clerical staff; sales staff; maintenance staff and transport drivers.
- Asset capital is the money invested in the plant and premises of a company.
- Working (trading) capital is the money required to buy raw materials, pay the wages and pay for services (e.g. gas, water, electricity, telephones) until the goods are completed and paid for by the customer.
- Profit is the money left over from the sale of manufactured goods after all the wages and expenses in their manufacture and in running the factory have been paid. The 'retained profits' after the payment of dividends are used to invest in research and development, new plant and to establish a reserve fund for expansion of the business and to keep the business going during times of recession.
- Quality can be defined as *fitness for purpose*.
- Quality assurance is the result of creating and maintaining an effective quality management system that insures all finished products achieve *fitness for purpose* and *conform* to the specifications agreed with the customer.
- ISO 9000 lays down the guidelines and rules for the setting up, accreditation and maintenance of a quality management system that is approved and recognized both nationally and internationally.
- Quality control is achieved by setting up a quality control system conforming with the requirements of ISO 9000 and which is supported by all persons working in a company from the chairman of the board down to the most junior recruit.

- Quality control points (points of inspection) must be inserted into the manufacturing chain at key points where further processing of a defective component would be a waste of time and money.
- Quality control points are usually inserted into the manufacturing system to check incoming raw materials and bought-in parts during material preparation, processing and after final assembly. The aim is always for a product to be *right first time* since the rejection and reworking of components result in reduced profits and loss of goodwill through late delivery.
- Quality indicators may be *variables* such as toleranced dimensions, or they may be *attributes* such as colours which can only be right or wrong.
- The outcomes of all sampling and inspection processes must be recorded so that, in the event of a product failing in service, the cause of the failure can be traced back to its source.

Evidence indicator activity 1.2

For a product of your own choosing:

1. Write a report that includes a general description of the key stages and scale of production with general notes explaining how changing the scale of production affects the organization of the manufacturing system.
2. Support the above written description with a *block diagram* illustrating the production system. Comment on the quality control points that are required to manufacture the product to specification.

1.3 Manufacturing organizations

1.3.1 Types of company

At the start of this chapter we examined the main elements of the manufacturing industry. Now we are going to look at the main elements of manufacturing companies. Companies will require many similar departmental functions no matter what product is being manufactured, for example:

- overall control (directors, proprietors, owners, etc.);
- production management;
- financial management;
- quality control;
- research and development;
- souring and procurement of raw materials and bought-in parts;
- marketing and sales;
- production workers;
- commercial services;
- stores and transport management.

Let's now look at the different types of company to be found within the manufacturing industries.

Private companies

These are companies which are wholly owned by a single person, a family or a small group of individuals. These persons own the companies outright and none of the shares is available for sale to the general public. Such companies can be further subdivided into proprietorships, partnerships and private companies with limited liability.

Public limited companies

These are large companies whose scale of operation and financing is beyond the resources of even the most wealthy private individuals. Such companies are funded by the sale of stocks and shares to the general public, and to the investment institutions such as the insurance companies. The stocks and shares in public limited companies (plc) are bought and sold through the stockbroking companies which, in turn, operate through the stock markets. Public companies pay interest on the money borrowed in this way. In the case of stock, a fixed rate of interest is paid. This may be less that the interest paid on ordinary shares and there is no capital growth. However, stocks have preference over shares in the event of the company failing and can be considered a safer investment.

The interest paid on shares is called a *dividend* and it is paid out of the profits of the company. When the profits go up, the dividend is raised and the shareholders get a higher rate of return on their investment. If profits fall, the dividend may have to be reduced. If the company prospers its shares will be in demand and, following the law of supply and demand, the price of the shares will rise and the shareholders will make a *capital gain* if they sell their shares at a higher price than what they paid for them originally.

Monopolies

These are companies which are free from competition because they are the only companies operating in a particular market. This is most likely due to the specialized nature of the service they offer and the high level of capital investment involved. A group of companies acting together to reduce competition and to keep prices and profits artificially high is called a *cartel*. However, the lack of competition can be unhealthy if the monopolies raise their prices and make excessive profits. If monopolies and cartels raise their prices too high, then it eventually becomes worthwhile for other companies to be set up despite the high level of investment involved. This breaks the monopoly and brings prices down to a more reasonable level.

Cooperatives

These are companies owned by the workforce, the management and, in some instances, by the customers as well. For all practical purposes cooperatives operate as limited companies with the shares owned exclusively by the members. The idea is to eliminate the profit element demanded by the more usual sources of capital funding. Any profits which are made are retained and reinvested in the business after an agreed dividend has been paid to the members. Such companies are registered under the Industrial and Provident Societies Act.

We have now discussed some of the different types of company to be found in the manufacturing industries, so in the next section we will look at the structure of a typical company.

Test your knowledge 1.15

1. Briefly explain the difference between private and public limited companies.

2. Some companies are classified as partnerships, proprietorships and cooperatives/operatives. Find out what these terms mean.

3. Briefly explain what is meant by:

    (a) a monopoly;
    (b) a cartel.

1.3.2 Management levels

The free flow of information between the departments of a manufacturing company is essential to the well-being of that company. In a small company consisting of the proprietor and two or three employees this is no problem. However, in large companies, information flow becomes more complex and companies have failed when the channels of information have broken down.

In a large company it is impossible for one person to solve all the problems, make all the decisions and carry out all the management duties. There comes a time in the development of the company when management has to be delegated. Fig. 1.42 shows a typical management structure and the interrelationships between the departments. This is the most common form of management 'tree', but there are others. Let's now look at the function of the various elements of the company.

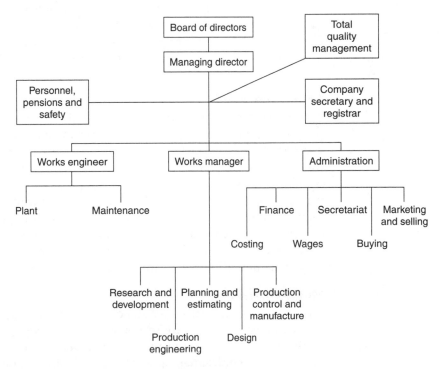

Figure 1.42 Company organization chart.

Board of directors

The directors are appointed to the board because of their technical and commercial experience, and/or because they have influential contacts which are beneficial to the company. They are usually professionally qualified in technology, accountancy, law, banking, insurance and marketing.

Figure 1.42 shows that the board of directors is at the top of the organizational tree of any limited company. The shareholders own the company and the directors are responsible to the shareholders for the successful running of the company. The directors:

- determine the policy of the company;
- authorise major investment projects;
- negotiate takeovers of and mergers with other companies;
- report to the shareholders;
- oversee the implementation of company policy;
- ensure that company policy is carried out efficiently.

In a public company the shareholders have the ultimate power to vote individual directors off the board. In fact they have the power to vote the whole board out of office if they are not satisfied with the running of the company. This rarely happens in practice.

Managing director

The directors, who are often part-time members of the board, communicate with the executive management of the company via a full-time managing director. The managing director is responsible for the overall running of the company and for the implementation of the board's policy. Alternatively, in some companies, this overall responsibility is undertaken by a chief executive.

Senior managers

These are managers who are highly qualified and members of the appropriate professions who head up the main departments of the company. They are responsible to the managing director or chief executive for implementing the policies of the board and for the day-to-day organization and efficient running of their departments. They are expected to show initiative in solving problems and to feed ideas back to the board via the managing director. They may, from time to time, be invited to present their requests and suggestions directly to the board and, in this way, become familiar with the way the board works. If they make a good impression over a period of time, they may be invited to join the board as a director when a suitable vacancy arises.

Middle managers

Many large departments need to be divided into sections supervised by the middle managers. They are responsible for the day-to-day running of their sections under the supervision of the senior manager. Generally they will have higher technician or equivalent qualifications. They will be responsible for ironing out the minor problems and difficulties that arise from day to day, and for ensuring that manufacturing proceeds smoothly and efficiently within the financial constraints laid down.

Supervisors, foremen, forewomen

These are persons who have considerable knowledge and skill at the 'shop floor' or 'office floor' level. They deal directly with the operatives, craftspersons and office juniors to ensure the smooth running of the company. They are responsible for implementing managing decisions and often advise the middle management on the practical difficulties that may arise from some decisions. The administrative breakdown of a large department is shown in Fig. 1.43

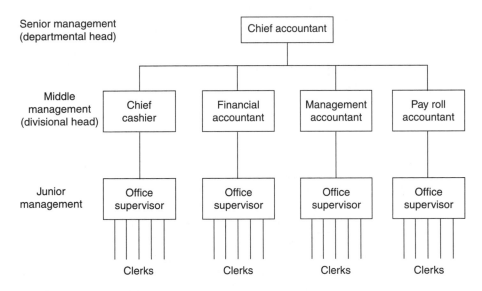

Figure 1.43 Departmental structure.

Test your knowledge 1.16

2. In the organization of a manufacturing company, describe briefly the roles of:

   (a) the board of directors;
   (b) the senior management;
   (c) the middle management;
   (d) the supervisors, foremen and forewomen.

2. Draw an organization diagram for the management structure of a small private company with only four managers reporting to the managing director.

1.3.3 Departmental functions

Financial control

Manufacturing companies exist for two purposes:

- To produce goods to satisfy a demand.
- To create wealth (make a profit) for the owners of the company.

To run at a profit there must be efficient financial control of the company. Overall control will be by the *financial director* who will be a chartered accountant. Assisting him will be the *chief accountant* who will also be a chartered accountant specializing in management accountancy. In turn, he will be responsible for the following accountancy functions:

- Payroll (wages) supervisor.
- Financial accountant (sales and purchase accounts).
- Management accountant (costing and estimating).
- Chief cashier (banking and insurance).

Commercial department

This department is responsible for all the non-technical aspects of the company. For example:

- stores;
- distribution of finished products;
- transport;
- export documentation;
- payment of wages and salaries;
- secretarial services.

This department will be headed by the *commercial manager*. In a small or medium sized company which does not have a financial director the commercial manager will also be responsible for the financial functions listed previously.

Stores are required for incoming goods, work in progress and for finished goods. Since the requirements for each of these responsibilities is different, separate stores are generally provided. The stores come under the control of the commercial manager. Basically, all stores require a controlled environment to maintain the materials and parts in good condition; a system for quickly locating the individual items and ensuring the stock is regularly rotated on a 'first-in, first-out basis'; and adequate room for mechanical handling. Some modern stores associated with the larger companies are fully automated so that the required items can be simply called up on a computer.

Many firms now insist on *just-in-time* (JIT) deliveries, that is, raw materials and bought-in parts are delivered only just before they are required. This not only saves on the cost and space but ensures the stock is fresh and in good condition. On the other hand, it often means increased storage of finished products so that the customer's delivery requirements can still be met even in the event of a production breakdown. This is essential to maintain 'goodwill'.

Quality control

Quality control used to be the role of the inspection department, which would inspect work in progress and the finished product and reject faulty items. This rejection represented considerable waste of time, materials and money. Nowadays quality control is a total company commitment from the board of directors to the most junior member of the company. The object is Total Quality Management (TQM) so that no defective work is produced in any department or by any person in the company. This prevents waste, ensures delivery on time and customer satisfaction. Many companies are now ISO 9000 approved and will only purchase goods and services from other companies which are also approved. This ensures product quality at all stages of manufacture. Because of the importance of quality control, the *quality manager* is responsible directly to the managing director and has oversight of the work of every department.

Company secretary and registrar's department

The company secretary often has a legal background and will also be a member of the Institute of Company Secretaries. The company secretary and his/her office are responsible for advising the managing director and the board on day-to-day legal matters; ensuring that the legal obligations of company legislation are met with; ensuring that the statutory documents are kept; and that the necessary returns are made to the appropriate government agencies. As registrar, he/she is responsible for the issue of share certificates, the keeping of the share register and the issuing of dividend cheques. Sometimes, in very large companies, the work of the registrar is subcontracted out to firms specializing in this work.

Personnel department

The *personnel manager* and his/her staff are responsible for implementing the employment policy of the company. As its name suggests, the personnel department is concerned with all aspects of the personnel employed by the company at all levels. Its main activities are as follows:

- appointments, discipline and dismissals;
- records of all employees at all levels;
- wage bargaining;
- employees' personal development;
- employees' personal performance;
- employees' welfare.

Training department

This is an important department (see also Section 1.3.3) and, in the larger companies, there is often a director with a special responsibility for training to support the training manager and his/her staff. Training covers a wide range of activities. For example:

- training graduate apprentices (technical and management);
- training commercial apprentices;
- training technician apprentices;
- training craft apprentices;
- training operatives;
- staff development programmes for existing employees at all levels particularly where new technology is concerned.

Works engineer's department

The works engineer's department (sometimes called the chief engineer's department) is concerned with:

- the installation and maintenance of the plant;
- the installation and maintenance of factory services such as gas water and electricity;
- the maintenance of the fabric of the premises.

The work's engineer will be assisted by *maintenance technicians* who have the technical knowledge to analyse and rectify problems that may occur in the sophisticated, computerized equipment used in most modern manufacturing companies. In addition, there will be *maintenance mechanics* who will carry out *preventive maintenance* such as routine lubrication and running repairs and adjustments as laid down in the operation manuals for the plant.

Sourcing and procurement (buying) department

The sourcing and procurement department has two responsibilities. First, to identify suitable suppliers of materials and parts that are required for manufacturing the company's products. Second, having found a suitable source for the materials and parts, to negotiate the purchase of such materials and parts. Heading this department is the *sourcing and procurement manager* who is responsible for ensuring that these materials and parts are obtained from reliable sources so that:

- a regular supply is maintained in order to avoid holding up production;
- the quality is of an acceptable standard;
- supplies are obtained at most advantageous prices;
- the sources are politically acceptable to the government of the day in the case of materials and parts imported from overseas.

Obviously this is a very important department. Failure to obtain supplies at competitive prices and at the right time to meet the demands of the production departments would cost the company dearly in lost production and excessive material costs.

Marketing and sales department

You may have a bright idea for a new product. You may impress yourself and your friends with its brilliance and quality. It will be *commercially useless* unless there is a *market* for it where it can be *sold at a profit* benefiting you as the designer and the firm that manufactures it for you. Therefore, before making any investment in design and development and in setting up a manufacturing plant, *market research* must be carried out.

Marketing and sales are often confused. Marketing is a *research activity*, its purpose is to find out what the customers want in time for new products to be developed and produced, and to organize a programme of advanced publicity to prepare the market for the new products. Once this has been done and the new products are available, the work of the sales team is to go out into the market place and obtain firm orders for the products. The *marketing division* of the department is also concerned with finding new markets and predicting the size of those markets. It must also be able to predict when the demand for a product is coming to an end so that the company is not left with surplus and unsaleable stocks on its shelves.

The *sales division* of the marketing and sales department is responsible for direct contact with the customer to achieve the sale of the goods made. This involves:

- initial contact with potential customers, either through direct selling, or selling through agencies and wholesale and retail outlets;
- negotiating contracts;
- providing technical information and back-up;
- providing feedback from the customer to the marketing department;
- acting as the company's interface with the public and presenting the 'image' that the company wishes to project.

The sales division is also responsible for the despatch of the finished goods and arranging suitable transport. This can vary depending upon the size, weight and quantity of the finished goods involved and the offloading facilities the customer may have. A company may have its own transport pool under the control of the transport manager for routine deliveries, or it may have to call in specialist contractors for the movement of very large and heavy loads.

The marketing and sales department must work closely with the *research and development department* and the *design department* to ensure that the company's products conform with the customer's requirements and to ensure that they keep ahead of the competition.

Manufacturing department

This department is headed by the work's manager. It can be subdivided into a number of divisions. Let's now consider the more important of these divisions.

Research and development. This division is responsible for the development of new products and services to satisfy future customer requirements as identified by the marketing and sales departments. It is also responsible for investigating new materials, new developments in technology and new manufacturing techniques, particularly if they lead to improved quality at reduced cost. All this is necessary to keep the company ahead of its competitors.

Design. This division is responsible for the design of both the new products and the tooling necessary for making those products. This not only involves the production of detail and assembly drawings but also the styling of the product so that it looks attractive to the purchaser. This is particularly important when the product is for sale to the general public, for example cars, washing machines and television sets. Companies often subcontract the styling to specialist studios.

The design department is responsible for formally setting down the work of the research and development department in the form of drawings and specifications that can be used for:

- tendering for contracts by the sales department;
- the manufacture of special tooling and measuring equipment for the manufacturing departments;
- the manufacture of any special maintenance equipment that may be required for the service department;
- the manufacture of the product itself.

Inspection and testing. This division is most important in maintaining product quality. It not only tests the finished products but also the incoming materials, components and subassemblies in order to ensure that the company's suppliers are maintaining the required standards. Further, it is no use waiting until the product is complete before finding out that it has been made from inferior materials. The inspection and testing division is also responsible for approving the prototypes of all new designs before they are put into production.

Production engineering. This division is concerned with the physical manufacture of the product. It must be brought into the design and development team at an early stage to advise on what can be manufactured within the limitations of the existing plant and what new plant may be required. Many a 'concept design' has had to undergo considerable modification before it could be made at a marketable price.

Production engineering is also concerned with the tooling and plant required for the manufacture of a company's products and with the manufactured quality of those products, and must also keep the senior management and board advised of new developments in plant and production methods so that a programme of plant and equipment investment keeps the company ahead of its competitors.

Production planning. This division is responsible for scheduling of all the details concerned with the manufacture of the product. This means planning:

- the time for the delivery of the raw materials and bought-in parts;
- the time for the manufacture of the components;
- the arrival of the parts and components at the assembly point as required;
- the completion of the job to meet the customer's requirements.

To achieve these objectives, the *production planning division* is responsible for scheduling (assigning) the workloads to the various operatives, machines and plant facilities.

Production Control. This division is responsible for implementing the schedules of the production planners and making them work. This involves:

- the preparation of all the necessary documentation for authorizing the manufacture of the product;
- controlling the movement of materials, manufactured components, bought-in parts and finished parts so that they arrive at the right place at the right time;

- progressing the work through the plant so that it is manufactured to schedule;
- taking corrective action if production falls behind schedule.

Product support. This division of the works manager's department has three areas of responsibility, all of which are directly concerned with customer satisfaction and in giving technical back-up to the sales staff.

- After-sales service and commissioning.
- Applications engineering, that is, providing advice to customers on the most effective way of using the company's products.
- Training the customer's workforce in the use of the company's products.

Test your knowledge 1.17

1. List the duties of the commercial manager in a small company that does not have a separate financial controller.

2. Describe the essential differences between the works engineer's department and the works manager's department.

3. Describe the difference between the company secretary and a private secretary to a manager.

4. Explain briefly what is meant by 'sourcing and procurement' and why this is such an important departmental function.

5. Explain briefly the essential differences between marketing and sales.

1.3.4 Skill and training requirements for work roles

We all receive a general education during the years we spend at school. This is intended to form a basis on which we can build further levels of education and training to suit specific work roles.

Professional qualifications

These require an honours degree plus specialist vocational training. At the highest levels a masters degree and/or a doctorate may be required. For example, a chartered accountant will start as a graduate trainee having first obtained an honours degree in, say, business studies. He or she will join a medium or large firm of chartered accountants which has suitable training schemes and will be trained 'on-the-job' while studying part-time to take the examinations of the Institute of Chartered Accountants. Several years of study and training are required and the standards are very high.

Similarly, chartered engineers are technologists who hold a degree in engineering and possibly higher degrees which they will have obtained through study at a university. This will have been supplemented by a post-graduate apprenticeship or pupillage. They will also have a number of years' experience in their chosen branch of the engineering profession in a responsible position.

Degrees

These are academic qualifications of the highest order. First degrees available at the:

- honours level (1st class, 2.1 and 2.2 classes, 3rd class);
- pass level;

- higher degrees such as 'masters' and 'doctorates' which can be awarded to holders of first degrees after further study and a successfully completed research project.

Most graduates are employed as much for their proven intellectual ability as for the actual knowledge acquired during their degree course. They are recruited into junior technological and management roles where they are given additional training and an opportunity to prove themselves for future promotion.

Diplomas and GNVQs

Traditionally, the Business and Technician Education Council issued diplomas in a range of subjects at the first year (F-level), the Ordinary National Diploma (OND-level and the Higher National Diploma (HND-level). These qualifications were intended for technician and incorporated engineers. The GNVQ (intermediate) has now virtually replaced the first year diploma, and the GNVQ (advanced) has now virtually replaced the ordinary national diploma (OND). At some future date two higher levels are to be introduced to run alongside and eventually replace the higher diplomas. GNVQs are issued by the City & Guilds, BTEC and the Royal Society of Arts (RSA). GNVQs (advanced) provide a path into manufacturing industry at the technician level or they can be used for entry into higher education to obtain a degree.

GCSE

General certificates of secondary education are awarded to school leavers at the end of their fifth form course, usually at the age of 16, provided they have reached a satisfactory standard in the GCSE examinations. An award is made for each subject passed. School leavers achieving a grade C pass or better in a minimum of English, mathematics, science and one other subject may elect to stay on to take a GCE A-level course or apply for a *modern apprenticeship* at the technician level. School leavers who only achieve D or E levels may also apply for a modern apprenticeship but at a craft level.

GCE A-level

The General Certificate of Education at the advanced (A)-level is also awarded on a subject by subject basis at various grades. 'A-levels' may be studied in the sixth form of a secondary school, in a sixth form college or in a college of further education. Passes at grade C or better provide entry for school leavers into higher education in order to obtain a degree. Alternatively, school leavers with good A-level qualifications may choose to enter manufacturing industry direct as junior management trainees. They will normally be encouraged and expected to improve there qualifications on a home study or part-time basis while in employment. Having looked at the various professional and academic qualifications available, it is now time to consider job related skill training.

NVQs/SVQs

National Vocational Qualifications (NVQs) and Scottish Vocational Qualifications (SVQs) are qualifications awarded for industrial skill training. Each of these qualifications is made up of a number of units. All the core units must be taken, to which can be added one or more optional units. Upon reaching the standard set out in a unit, a certificate will be awarded and credited towards the full NVQ. When sufficient units have been obtained an NVQ will be awarded. There are five levels in the NVQ framework with level 1 as the lowest and level 5 as the highest award.

The skill training necessary to obtain these qualifications can be gained at work, in skill training centres or at colleges of further education. Assessment can take place at work and credit can be given immediately for skills already achieved. NVQs and SVQs are intended to provide a path for existing employees to obtain promotion and for school leavers to obtain employment.

Modern apprenticeships

These are based on achieved standards of performance instead of being based on 'time served' as in traditional apprenticeships. They are open to young males and females between the ages of 16 and 17 who have achieved appropriate GCSE subjects with an adequate level of pass as stated above. The aim of modern apprenticeships is to provide manufacturing industry with young people possessing high quality skills and potential for further development. At the end of their apprenticeships they should have achieved NVQs at level 3 and they may be recommended to progress to higher education. They will also have studied the necessary underpinning background theory to support their practical skill training.

Company training certificates

Employees at every level in a company represent an expensive and potentially valuable asset. Therefore many companies recognize the importance for staff development as a means of enabling staff to achieve their full potential and also helping to provide job satisfaction. To attain this end many companies provide courses tailored to their particular requirements, either through their own personnel department/staff training school or through the services of outside training consultants and organizations.

Normally such courses are skill orientated, with a minimum academic input, and are highly specialized–for example, introducing staff to new computer software and working practices. Depending upon the depth of training required, courses may last a day, a week or even longer. To avoid undue disruption longer courses may be spread over, say, a year on a one day per week basis. It depends how quickly the new knowledge and skills need to be acquired. On successful completion of the course a company certificate is awarded and the personnel file of the recipient is marked up accordingly. This will be taken into account when promotion is applied for.

Having considered academic qualifications and skill training, we now have to see how these qualifications and skills can be applied to various roles in the manufacturing industry. Let's first consider the administrative staff.

Senior managers

Senior managers will be professionally qualified with years of industrial experience behind them and a proven track record. They will be chartered engineers, chartered accountants or hold higher degrees from the specialist business and management schools of the leading universities.

Middle managers

These are professionally trained people who are still gaining management experience. They will usually be university graduates in disciplines such a manufacturing and design engineering, chemical engineering, accountancy or law. They will also have passed the qualifying examinations of their appropriate professional institution. They will assist senior managers by heading the divisions within departments and will need to be continually updating their knowledge of changes in technology, management practices and company structures.

Junior managers

These will have graduate qualifications or higher national diplomas but will not yet have attained professional status. They will be assisting the middle management and will be gaining management skills both on the job and through further study. They may also be gaining 'shop floor' and 'office floor' experience, particularly in such fields as interpersonal skills and 'man-management', as assistants to supervisors.

Supervisors

Supervisors are well-qualified persons with higher C&G, BTEC or RSA qualifications. Mainly, however, they will have years of experience in direct control of the office clerical staff and the manufacturing operatives and craftspersons. They will most likely have a Certificate in Supervisory Management. Company training packages have to be tailored to suit the requirements of each supervisory post since, even in the same company, the work of supervisors can be very different from department to department.

Having looked at the roles of the administrators (managers and supervisors), let's now look at the roles of the technologists, technicians, clerical workers and production workers.

Technologists

Technologists will be university graduates with several years' experience gained during a *post-graduate apprenticeship*. This will have enabled them to gain experience in all the major departments of the company. Their degrees will depend upon the products being manufactured. For example, they may be graduate engineers, chemists, brewers or food technologists, depending upon the industry in which they are working. They will be employed in such activities as:

- research and development;
- design;
- process control;
- manufacturing;
- maintenance.

Incorporated engineer

Because of the confusion between the *engineering technicians* and the higher level, previously known as *technician engineers*, the career title of *incorporated engineer* has been introduced in place of technician engineers. Incorporated engineers will have studied to at least the Higher National Diploma level. Increasingly employers are requiring an engineering degree. Training is usually on the job under the mentorship of a professional engineer, combined with study for professional qualifications. Incorporated engineers are often team leaders, supervising technicians and craftspersons. They use their knowledge and training to solve all sorts of engineering problems, in both production and design.

Technicians

Technicians cover a variety of jobs. They will have studied to Ordinary or Higher National Diploma levels. They may work in drawing and design offices, produce and develop prototypes, work in test laboratories, be responsible for quality control and production control, or they may supervise the maintenance of highly sophisticated computer controlled machine tools and robots. In a small firm they may have to work independently, relying on their own skill and initiative to analyse and solve problems. However, in a large firm they usually work under the direct supervision of a professional engineer and are often part of a team.

Production and clerical personnel

Craftspersons. Craftspersons are expected to develop high levels of skill and be capable of working with general purpose machine tools and hand (bench) tools. They will work with a minimum of supervision and be expected to turn out work of high precision. In view of the type of training to be undertaken, many employers use their own aptitude tests for manual dexterity. Training will usually follow or be similar to the structure of the modern apprenticeship.

Operators/assemblers. Operators work machines in an engineering factory: assemblers put things together. The machines will have been set by suitably qualified setters and the operator will merely operate the machine and/or load and unload it. In the clothing and related industries they will usually be known as machinists. Operators are only trained on a 'need-to-know' basis and are responsible to their supervisors for the quality of their work. Although for this type of work school leaving qualifications are not essential, most employers expect to see some evidence of a reasonable level of literacy and numeracy. This is required in order to understand and benefit from the training given and, most important, an understanding of the hazards involved and the safety procedures applicable to the process.

Clerical support staff. These range from highly qualified and experienced personal secretaries and personal assistants (PAs) to the senior management down to the office juniors recruited straight from school. The clerical support staff include:

- account clerks;
- computer operators;

- print room operatives;
- switchboard operators;
- post room operatives;
- receptionists.

Miscellaneous

In addition to the work roles just discussed, there are a whole range of jobs important to the manufacturing process but not necessarily requiring formal academic study. Frequently NVQs will be encouraged for purposes of personal development and job satisfaction. These work roles include:

- machinists;
- clerical workers;
- print room and post room operatives;
- shop floor operatives;
- stores and warehouse personnel;
- light and heavy goods vehicle drivers;
- labourers and cleaners.

Test your knowledge 1.18

1. Briefly describe the role of, and the qualifications required by, the following company personnel.

   (a) chief financial controller;
   (b) research chemist;
   (c) master brewer;
   (d) production manager;
   (e) personal secretary;
   (f) maintenance technician in an engineering factory;
   (g) machinist in a clothing firm;
   (h) wages clerk.

2. On leaving school, your ambition is to enrol on a modern apprenticeship in order to become an engineering technician.

   (a) Find out what school leaving qualifications you will require in addition to your GNVQ in manufacturing.
   (b) Write a brief report on what typical theoretical and practical skills you can expect to gain from such an apprenticeship.

Key Notes 1.3

- Small companies can be partnerships, proprietorships or *private limited companies*, financed by the owners personally, with or without the help of their banks.
- Large companies are usually *public limited companies* financed by selling shares on the stock markets. These shares may be bought by private individuals or by financial institutions such as insurance companies.
- In a public limited company the shareholders own the company and appoint a board of directors to run the company on their behalf. The directors are responsible to, and must report to, the shareholders at least once a year.
- Cooperatives are owned by the members who may be the workforce, the management and, in some instances, the customers as well. For all practical purposes they operate as limited companies with the shares held by the members who receive a dividend.
- A *monopoly* is created when a single company dominates a market without competition so that it can set whatever prices and trading conditions it likes for its products.

- A *cartel* is created when a number of companies act together to create a monopoly–for example, OPEC which controls most of the world's major oil producing companies.
- The managing director (or the chief executive) is the link between the board of directors who set the policy of the company and the senior departmental managers who implement the policy decisions of the board.
- Senior managers are heads of departments and are assisted by middle managers who head the divisions within a department. Below the middle managers are the supervisors, foremen, forewomen and charge hands.
- The major departmental functions in a company (arranged alphabetically) are: commercial; financial; maintenance; manufacturing; marketing and sales; personnel and training; production control; production engineering; quality control; research and development; sourcing and procurement.

Evidence indicator activity 1.3

1. For a company of your choice, write a report, in general terms:

   - describing the departmental functions within that company;
   - identifying the key features of the main work roles for each of the departmental functions set out in (a).

2. As part of the above report produce:

   - a diagrammatic illustration of the main work roles.
   - a diagram illustrating how the departments interact during the manufacture of the company's products.

3. Produce a summary of the roles and responsibilities within the departments using organizational charts, and identify the main skill and training requirements for each role.

1.4 Environmental effects of production processes

Production processes have always affected the environment. Originally the UK was heavily forested. As communities began to develop, areas of forest were felled to make room for hamlets to be built and land to be farmed. The trees that were felled provided timber for building and fuel for heating. Later, trees were felled to source the raw material for charcoal burning. Large swathes of mature oak forests throughout the land were felled to provide the wood to build the ships for the much enlarged Royal Navy from the Tudor period in our history until warships built from metal became the norm. The landscape was changed for ever and continues to change as industrial and domestic requirements make increasing demands on our remaining natural resources. Nowadays we appreciate the damage done to our environment by industry in the past. Therefore, modern industry is much more environmentally conscious.

BS EN ISO 14001 (Environmental Management Systems) is the recognized standard for the manufacturing and service sectors of British industries with regard to the management of environment issues. Companies can obtain certification in the same manner as ISO 9000 quality assurance. The standard covers all the areas shown below.

Requirements

1. The organization shall establish and maintain an environmental management system.
2. Environmental policy
   - Products and services
   - Commitment to improvement
   - Commitment to comply with relevant regulations
   - Framework for reviewing objectives
   - A policy available to the public
3. Planning
   - Legal requirements
   - Objectives and targets
   - Environmental management programme
4. Implementation and operation
   - Structure and programme
   - Training
   - Communication
   - Documentation
   - Operation control
   - Emergency response
5. Checking and corrective action
   - Monitoring and measurement
   - Non conformance, corrective and preventive action
   - Records
   - Environmental management system audit
6. Management Review

1.4.1 Environmental impact

We can view this in a number of different ways. For example, the effects of manufacturing processes on:

- quality of life in terms of human beings, natural resources and buildings;
- local economy;
- renewability of resources;
- local geography;
- pollution.

Let's consider these in turn.

Quality of life (human)

The goods produced by the production processes of the manufacturing industries have greatly improved the quality of life for the majority of people. Such goods as cars, motor cycles, fabrics, clothes, computers, television sets, food, medical equipment and medicines are all readily available. Public transport by buses, trains, aircraft and ships has shrunk the world and made possible frequent and regular journeys that, a few generations ago, would have seemed impossible except for the very wealthy.

On the downside, some of these processes have resulted in pollution of the environment that has caused serious distress and even illness for those who live near to the plants concerned. Pollution and waste disposal will be considered in greater detail later in this chapter.

Quality of life (natural)

Natural resources can be subdivided into:

- those that are irreplaceable, such as fossil fuels and metallic ores, together with mineral deposits such as rocks, sand, gravel and the raw materials for ceramic products;
- those that are replaceable by planting and breeding.

Some examples are shown in Fig. 1.44. Unfortunately even the use of natural resources has its downside as intensive farming results in the destruction of hedgerows and the draining of wetlands with the inevitable loss off wildlife habitats. Also the increasing use of nitrate fertilizers, pesticides and insecticides has resulted in dangerous pollutants getting into the food chain and drinking water. A balance needs to be maintained between the economics of production and the quality of life.

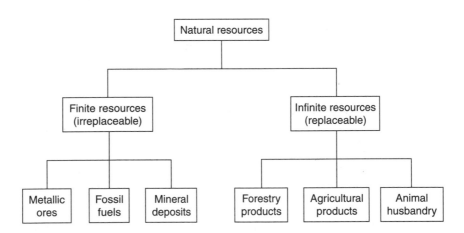

Figure 1.44   Natural resources.

Quality of life (built)

Construction has changed the face of every country far more than any other human activity. The products of the construction industry have resulted in better housing, more convenient and pleasant working environments and better road and rail networks leading to improved transport systems. The downside is the destruction caused by quarrying and mining for the raw materials and the pollution caused by the conversion of raw materials into building materials. On the other hand, the waste products of coal fired power stations provide the raw materials (ash) for manufacturing insulation blocks and plaster (gypsum) for facing the interior walls of buildings. The recycling of this waste from our power stations prevents it being added to the ever increasing problems and cost of waste disposal, see Section 1.4.2.

Local economy

Manufacturing creates the wealth that sustains local and national communities. On a local level it provides employment for the residents of the local community. Indirectly it also creates employment possibilities by creating the wealth and need for service industries such as shops, schools, recreation centres and public transport, as indicated in Fig. 1.45.

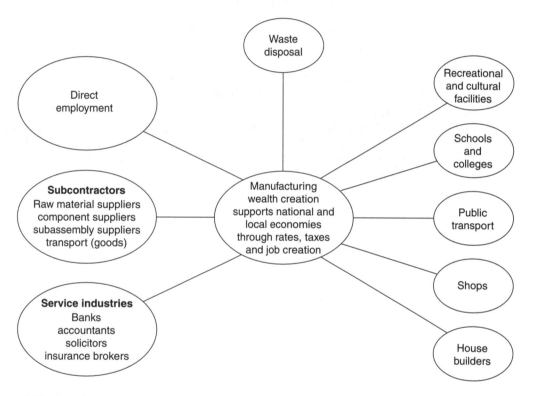

Figure 1.45 Local economy.

Renewability of resources (natural)

Only natural resources that can be grown or bred are renewable. We cannot restore mineral resources. You *cannot* replace the fossil fuels and minerals once they have been extracted. You *can*, however, grow cereal and vegetable crops, tea and coffee, and replant forests and rubber tree plantations. Every effort should be made to use only timber from quick growing softwood species and reduce the destruction of the slow growing hardwood forests of the equatorial regions. These latter forests are not only virtually impossible to replace but provide important wildlife habitats and have a profound effect on the world's weather systems.

Renewable resources from the animal world include meat, hides for leather, milk and eggs. Fish farming is still in its infancy and will never compete in volume with the self-renewing natural resources in the seas, providing the fishing industry is sensible in its extraction policies and does not allow greed and indifference to destroy this valuable source of food.

Local geography

Mankind has constantly changed the local geography from the times of the first settlements to the great cities and conurbations of today. The change continues as our requirements change and new construction materials, technologies and techniques become available. Some changes have ignored the forces of nature and resulted in mini-disasters. For example, stormwater that was once allowed to soak into the land locally to restore underground supplies (aquifers) is now ducted to our rivers which become overwhelmed and flood. This causes destruction of property and distress to those living in low-lying areas liable to flooding.

Local geography is also subject to continual change as a result of the continually changing requirements of industry. For example, the steady decline of the heavy, 'smoke-stack' industries such as coalmining, shipbuilding and iron and steel threatened to turn many areas of the country into industrial wastelands and to destroy the local economies dependent on those industries. Much effort has been put into rescuing these areas economically by attracting local and overseas investment in high technology industries such as the manufacture of computers and their associated components. The development of attractively landscaped, modern industrial estates for light industry is replacing the 'dark satanic mills' of the Victorian industrial revolution.

Pollution

Officially, a pollutant is a substance which is present at concentrations that cause distress, harm or exceed an environmental quality standard. Such pollutants are mainly carried by air or by water as shown in Fig. 1.46. Let's now consider each category in more detail.

Airborne pollution

Airborne pollutants can enter the atmosphere directly from the production process or they can be the result of chemical reactions in the atmosphere.

Combustion. The burning of wood and fossil fuels is a major source of atmospheric pollution. These pollutants are smoke particles and gases such as sulphur dioxides and various nitrogen oxides, together with carbon monoxide.

- Smoke particles can consist of burnt fuel particles such as soot, unburned fuel particles and fine ash, and traces of various other solids.
- Sulphur dioxide is a chemical compound formed by the sulphur in the fuel reacting with the oxygen from the air required to burn the fuel. The sulphur dioxide gas reacts with the moisture in the atmosphere to produce the so-called and damaging 'acid rain'.
- Nitrogen oxides result from reactions between the oxygen and the nitrogen in the air at the high temperatures in modern furnaces. This

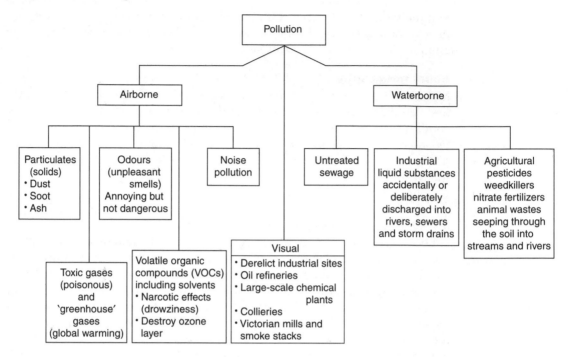

Figure 1.46 Sources of pollution.

gas also contributes toward the 'acid rain' and ground level ozone. This ozone is an irritant for persons subject to respiratory diseases such as asthma. It can also affect the growth of sensitive crops.

- Carbon dioxide is the major product of the combustion process. It is a chemical compound formed between the carbon in the fuel and the oxygen in the air when combustion is complete. It is the 'greenhouse gas' most responsible for global warming.
- Carbon monoxide gas is the result of incomplete combustion between the carbon in the fuel and the oxygen in the air. It is a toxic substance. Toxic substances are poisons. In small doses it leads to dizziness, lack of physical coordination, impaired vision and impaired judgement. In larger doses it kills.
- Dust particles such as arsenic, lead, zinc, iron and other metals can get into the atmosphere from the combustion of fuels and the metal extraction, manufacturing and plating processes. In sufficient concentrations these can have long-term effects on people's health if inhaled over a period of time.
- Volatile organic compounds (VOCs) are released from a variety of toxic chemicals used in the plastics and adhesives industry, from dry cleaning metal degreasing plants. Such substances readily evaporate into the atmosphere and, in sufficient concentrations, can have narcotic (anaesthetic) and toxic (poisonous) effects. They also combine with the nitrogen oxides and accelerate the formation of ground level ozone. Continual exposure to VOCs can cause serious and permanent damage to health.

Noise. This is also a form of airborne pollution since sound waves travel through the air. It can irritate and cause loss of concentration. It represents wasted energy and prolonged exposure to noise can cause permanent damage to the nervous system and also permanent deafness. Nearby neighbours of noisy plants and processes can have their quality of life destroyed. Careful design of plant and buildings and the erection of noise barriers can all lead to a reduction of this type of pollution.

Odour pollution. Odours are also carried through the air. Although noxious (unpleasant) they are mainly an annoyance and unlikely to have any long standing medical effect. Some typical causes of unpleasant smells are:

- painting and enamelling;
- factory farming;
- food processing;
- chemical manufacture;
- plastic moulding;
- brewing.

Waterborne pollution

This results from accidental and deliberate contamination of rivers and other water sources. It can occur from accidental spillage or from the deliberate discharge of industrial pollutants into rivers and storm drains. It can also occur by the leeching (seeping) of pollutants through the soil. This occurs when farmers use excessive amounts of nitrate fertilizers, insecticides, pesticides and animal manures on their land. It can also occur when toxic substances are dumped in landfill sites that have not been properly engineered and managed. The main danger to humans is when such pollutants find their way into drinking water supplies as they are difficult and costly to neutralize and/or remove. Untreated, such pollution is also a hazard to freshwater ecosystems and saltwater ecosystems.

Visual pollution

The old industries were notorious for visual pollution.

- The spoil heaps and pit-head gear of coalmines.
- The blastfurnaces of iron works.
- The smokestacks of the woollen and cotton mills.

These are just a few examples of the traditional visual pollution of an earlier age which can still be seen in some parts of the country. Nowadays industrial developments are expected to be landscaped to reduce the visual impact. Also firms engaged in opencast mining and mineral extraction are expected to restore the site to an acceptable condition once the minerals are worked out and extraction ceases.

Test your knowledge 1.19

1. Explain briefly why the UK is no longer heavily forested and in what way manufacturing over the ages has been responsible for this deforestation.

2. Write a brief report on how manufacturing has affected your local community. Your report should include the effect local manufacturing industry has had on the quality of life, the local economy, renewable resources, local geography and pollution.

3. With the aid of examples of your own choice, briefly explain what is meant by the following terms relating to pollution.

   (a) Airborne pollution.
   (b) Waterborne pollution.
   (c) Visual pollution.

1.4.2 Waste disposal

Waste can be defined as anything that has no further use. It is something that has to be disposed of with as little impact on the environment as possible. How waste is disposed of depends on what it is. Remember that, in some instances, the waste from one industry can be the raw material for another. Fig. 1.47 shows the different categories of waste.

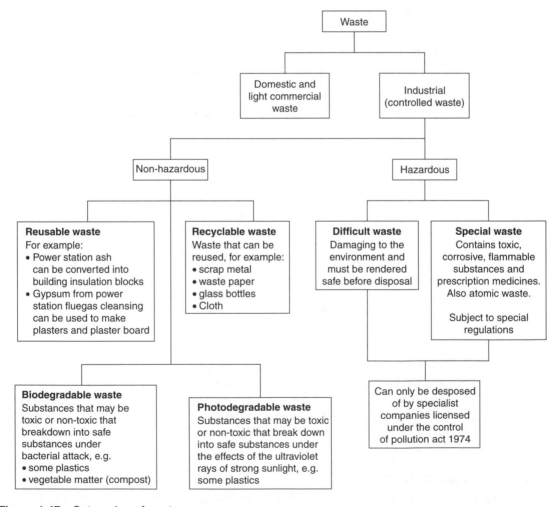

Figure 1.47 Categories of waste.

Industrial waste

Industrial waste is classified as *controlled waste* under the Control of Pollution Act of 1974 and it can be divided into:

- Hazardous waste.
- Non-hazardous waste.

Metal manufacturers, chemical manufacturers, abattoirs and food processing plants, and power stations are all examples of major contributors to the problem of industrial waste disposal. Much of the waste, such as power station ash, is *non-hazardous* and is only of inconvenience because of the immense amount produced. Other waste can be detrimental to the environment. This is classified as *hazardous waste*. Hazardous waste can be subdivided into two categories:

Difficult waste. This is potentially damaging to the environment and needs to be treated to render it safe before discharging it into the rivers or the sea in the case of liquids or landfill sites in the case of solids. An example is the disposal of lubricating and cutting oils.

Special waste. This is waste containing toxic and other dangerous substances. It is the subject of additional regulations. Examples of special waste include poisonous substances, corrosive substances (acids), flammable substances and prescription-only medicines.

Both difficult waste and special waste can only be disposed of by specialist waste disposal companies which have been licensed under the Control of Pollution Act 1974.

Reusable waste

This is the waste from one industry that can be the raw material of another – for example, the waste ash from a power station can be used to produce building blocks with good load bearing and thermal insulation properties for the construction industry. Power stations also take care to remove the sulphur from the flue gases by allowing it to react with limestone to form *gypsum* (calcium sulphate) before the gases leave the smoke stacks. The gypsum is a raw material for the manufacture of cements, plaster and plasterboards for the building industry. Any surplus gypsum is welcomed by the operators of landfill disposal sites as it is non-toxic and stable. Another example of reusable waste is the spent hops from the brewing process. The spent hops can be used to make a compost much in demand by mushroom growers.

Recyclable waste

This is waste that can be collected, reprocessed and reused as a raw material for manufacturing. This helps to save dwindling resources of non-replaceable raw materials. Examples of such recyclable waste are:

- Scrap metal such as cast iron, steel, aluminium and copper that can be melted down and reused.

- Waste paper and cardboard that can be shredded, pulped and converted back into paper and board again for reuse.
- Waste glass (shatter) that can be added to new molten glass and reused. Bottles, such as milk bottles, can be returned to the dairy for sterilization and refilling.
- Textile offcuts and rags can be either recycled by the textile industry to produce a cloth called 'shoddy' or used as cleaning rags (wipers) in industry.
- Waste plastics, solvents, oils and rubber can be chemically treated to make them suitable for reuse.

Toxic waste

This is waste material that is harmful to human beings and the environmental ecosystems. Some examples are the heavy metals such as lead and mercury. These are used in batteries and are systemic poisons affecting the nervous system. Other toxic substances include:

- Metals such as cadmium and arsenic.
- Metals and chemicals used in electroplating processes.
- Chemicals used in the preservation of timber.
- Solvents used in the manufacture of paints and in the degreasing of metals prior to the application of finishes.
- Pesticides, insecticides, asbestos products and many chemicals used in the manufacture of plastic materials.

None of these can be got rid of just by dumping. They must be collected and made safe by licensed specialists before final disposal. Radioactive waste is a special case and the difficulties of its disposal are widely discussed in the press.

Non-toxic waste

Non-toxic waste is neither fatal nor harmful to humans or the environmental ecosystems. Examples of non-toxic waste are the ash and gypsum from power stations already mentioned, builders' rubble, wood, ferrous metals, aluminium and its alloys, copper and its alloys, and garden waste. However, vegetable matter dumped in landfill sites decomposes (rots). In the process of decomposing, it shrinks and can cause subsidence and instability. It also gives off methane gas which is flammable. For this reason, landfill sites must be properly ventilated. Wherever possible vegetable matter should be recycled by composting.

Degradable waste

Degradable waste can be toxic or non-toxic and decomposes after disposal by one of two processes.

- *Biodegradable waste* is broken down by bacterial attack—for example, vegetable matter is broken down in this manner to provide a recyclable compost.

- *Photodegradable* waste is broken down by the action of the ultraviolet rays in sunlight. Many plastics break down in this way. However, plastic window frames are specially compounded (UV resistant) to prevent this happening. Other plastics are deliberately made to be either biodegradable or photodegradable so that they will decompose harmlessly after disposal.

Non-degradable waste

This is toxic or non-toxic materials that do not readily decompose–for example, builders' rubble and metals. Given sufficient time most materials deteriorate if left exposed to the environment.

Byproducts

Byproducts are not waste as defined above, but *secondary products* of a manufacturing process that can be sold and which contribute to the profitability of the company. For example, when crude oil is refined to extract petrol many other products are also produced as a result of the refinement process. These include diesel oil (derv), lubricating oil and chemicals used by the plastics industry. In this case, petrol is the main product and the rest are the byproducts. Without a market for the byproducts, petrol would be very much more expensive.

Waste disposal

The five most common methods of waste disposal used in the UK are:

- landfill sites;
- incineration;
- disposal at sea;
- disposal into water courses (rivers);
- composting.

Landfill sites

Most waste solids in the UK are deposited in landfill sites. In fact 85% of all industrial waste is disposed of in this manner. Care has to be taken that such waste is non-toxic or has been treated to render it harmless. The scale of disposal in landfill sites is enormous. It has been estimated that if, in one year, all this waste was loaded into heavy lorries they would fill a six lane motorway from London to Tokyo! The Waste Regulatory Authority (WRA) insists on all landfill sites being properly engineered and managed.

- An engineered pit lining is constructed to seal the waste from surrounding rock, soil strata and water table.
- Water entering the site must be contained within the site. Capping systems and small working faces restrict the ingress of water.
- Rubbish must be deposited in consistent, even layers to strict engineering procedures to ensure safe decomposition and a stable body of refuse.
- The decomposing body of waste can generate large volumes of landfill gas (LFG) and noxious liquids (leachate). The site must be regularly checked and inspected for gas migration and ground water quality.

- Currently, 70% of LFG (mostly methane) is allowed to escape into the atmosphere. This is not good for the environment since LFG contains greenhouse gases and it is also highly flammable. If it collects in pockets it can be explosive. The rest is either flared off or collected for electricity generation. Of the 600 megawatts (MW) total generated in the UK only 32 MW are produced from landfill waste. However, this figure is rising.
- Landfill operators must not only provide reassurance of minimum impact on local communities during the site's productive life, but for many years after it has been filled.
- Filled sites offer opportunities for landscaping and redevelopment. Restoration is now an important part of landfill management, hence the need to ensure that such sites are left in a stable and safe condition.

Incineration

This is used not only to reduce the weight of waste by up to 75% and its volume by up to 90%, it also sterilizes the waste and often renders toxic and noxious substances harmless. Incinerator ash can then be either disposed of at landfill sites or used in the production of horticultural composts. Refuse that has been incinerated will not decompose and produce LFG. This improves the safety and stability of any landfill site where it is finally deposited.

Disposal at sea and in water courses

At one time waste of all kinds, toxic and non-toxic, was dumped at sea and in the estuaries of our larger rivers. As the volume of such disposal increased and the substances became more toxic, the potential and actual ecological damage to the marine environment led to the need for close control. Nowadays only safe, non-toxic solids can be dumped at sea and the quantities involved are relatively small compared with those deposited in landfill sites. Such solids must be able to sink so that they do not wash up on the shores of the UK (e.g. rubble and power station ash) or it must be biodegradable in a marine environment. It must not be capable of affecting or infecting any form of marine life.

Liquid effluents have long been discharged into rivers, estuaries and around coasts without treatment. Apart from such industrial effluents, sewage has been the main cause of contamination. Nowadays the discharge of untreated effluents is not allowed. All liquids so discharged come under the jurisdiction of the officers of the Industrial Pollution Control (IPC) department of the former National Rivers Authority (NRA). The NRA has recently been absorbed into the Environment Agency (EA). This organization provides *control through consent to discharge*. Their requirements include:

- temperature control of the effluent;
- control of the rate of effluent discharged per hour;
- control of the pH value of the discharge;
- the sediment content, heavy metal content and toxicity;
- satisfying the bio-oxygen requirements of the ecosystem of the watercourse concerned.

These same requirements are now also applied to liquid effluents released into river estuaries and coastal waters.

Composting

Waste vegetable matter from food processing can be usefully turned into horticultural and agricultural composts and help to save the world's dwindling reserves of peat. It has already been mentioned that spent hops from the brewing process are used as the basis for the compost used by mushroom growers.

Test your knowledge 1.20

1. State the two main classifications of industrial (controlled) waste and give an example of each.

2. Briefly explain the difference between 'difficult waste' and 'special waste'.

3. Briefly explain the difference between 'reusable waste' and 'recyclable waste'.

4. Give an example of each of the following, with reasons for your choice:

   (a) toxic waste;
   (b) non-toxic waste;
   (c) degradable waste;
   (d) byproducts.

5. List the main methods of waste disposal and give an example where each method of disposal can be safely used.

1.4.3 Energy sources

Figure 1.48 shows how the energy sources available may be categorized. There is no difference between the sources available to manufacturing industry and the resources available generally.

- *Finite energy sources* are also known as *non-renewable energy sources*. Once used they can never be replaced.
- *Infinite energy sources* are also known as *renewable energy sources*. They can be replaced or reused over and over again.

Finite energy sources

Let's first consider examples of the *finite energy sources* in more detail. These are all fossil fuels such as coal, oil and natural gas. Coal is the fossilized remains of plant life that grew millions of years ago in prehistoric times. Oil and natural gas come from the fossilized remains of minute marine creatures known as *plankton*.

The chemical energy stored in fossil fuels is released by burning the fuel to convert this chemical energy into heat energy capable of doing work. The burning (combustion) of fossil fuels not only releases the chemical energy in the form of heat energy, it also releases undesirable *products of combustion* such as smoke and gases into the atmosphere causing pollution.

Air pollution

Earlier in this chapter it was stated that when fossil fuels are burnt, smoke (dust and unburned or partially burnt particles of fuel), sulphur compounds, nitrogen compounds and carbon compounds are released into the atmosphere. These products of combustion can cause harm to plant life

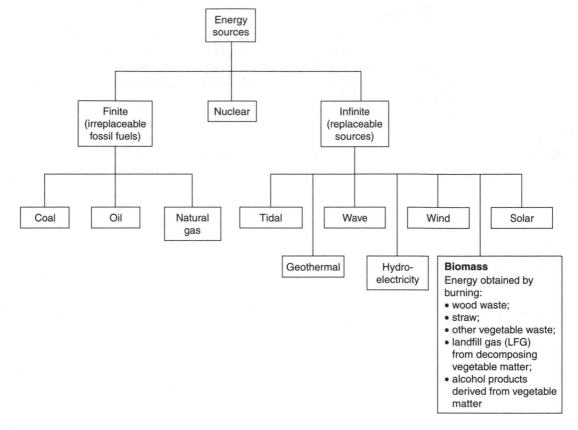

Figure 1.48 Energy sources.

and buildings and can cause distressing respiratory problems for animals and human beings alike.

Acid rain

As mentioned earlier in this chapter, sulphur is present in varying amounts in most fossil fuels. When the fuel burns it combines with the oxygen of the air and so does the sulphur to form oxides of sulphur. These oxides of sulphur are gases that rise into the upper atmosphere were it combines with any water vapour present to form dilute sulphurous and sulphuric acids. These acids are carried back to earth in the form of acid rain deposition to cause serious pollution. The acid rain deposition can get into trees and plants through the soil and through their foliage. This results in restricted growth of plant life. In sufficient concentration, acid rain:

- can kill plant life and, even in lesser concentrations, can weaken plant life to the extent that it becomes susceptible to disease;
- can cause damage to buildings by chemically attacking building materials such as stone, concrete and metal;
- can affect the pH value of lakes, rivers and streams causing depletion of fresh fish stocks and related ecological damage.

Nitrogen compounds

Air is needed to burn fossil fuels and four-fifths of air is nitrogen. Normally nitrogen is an essential and harmless constituent of the atmosphere. Modern furnaces and internal combustion engines (as used in vehicles) tend to operate at higher temperatures than in the past in order to be more efficient in the amount of fuel used. Unfortunately these higher temperatures enable the nitrogen to combine with the oxygen in the air to form various nitrogen–oxygen compounds. Like the sulphur compounds, described previously, these oxides of nitrogen (NOx) combine with the moisture in the atmosphere to form acid rain. Worse, under suitable atmospheric conditions (strong sunlight and high temperatures), these nitrogen compounds can cause ground level ozone and smog. These conditions are most frequently met with in large cities where large volumes of NOx gases are emitted from vehicle engines. Ground level ozone can be distressing and even dangerous to people suffering from respiratory diseases such as asthma. It also affects people's eyes causing blurred vision and acute discomfort.

Greenhouse gases

Fossil fuels are hydrocarbons, that is, they are compounds of hydrogen and carbon. When carbon burns in air the carbon combines with the oxygen and forms the gas called carbon dioxide. This is the main constituent of the products of combustion produced by burning fossil fuels. If the fuel is not burnt efficiently then some of the toxic and flammable gas carbon monoxide will also be present. Both these gases are *greenhouse gases*. Other greenhouse gases are ozone, methane and nitrous oxides. These gases form a belt high in the atmosphere and behave like the glass roof of a greenhouse, hence their name. In moderation they are beneficial and without them life could not exist as we know it.

- They allow the warming rays of the sun to reach the surface of the Earth.
- They filter out the more harmful radiations such as ultraviolet rays.
- They trap the heat in the atmosphere at night and prevent excessive cooling.
- They help to maintain an equitable temperature range in which life as we know it can exist.

However, if you have too many blankets on your bed the balance between the heat being radiated from your body and the heat escaping through the bedclothes is upset and you become uncomfortably hot. The same thing happens if there is an excess of greenhouse gases. The Earth receives more heat energy from the sun than is lost back into space. This causes the average atmospheric temperature to rise and we have *global warming*. If we continue to burn fossil fuels at the present rate then it has been estimated that the global temperature will rise at approximately 0.5°C per decade. This does not seem much, but already there are signs that the polar icecaps are beginning to melt. If this happened on a larger scale the increased volume of water in the seas and oceans of the world would result in the flooding of vast areas of low-lying land. Also the release of such a large volume of very cold water into the oceans would upset the circulation of the ocean currents and this, in turn, would change the whole climatic system.

Tables 1.13, 1.14 and 1.15 summarize the advantages and disadvantages of burning fossil fuels.

Table 1.13 Advantages and disadvantages of coal as an energy source

Advantages	Disadvantages

• The most widely available fuel worldwide and in mainland UK
• Lowest cost fuel
• Techniques for its extraction and use are well established
• Its extraction can provide local UK employment
• Most widely used fuel in the UK for electricity generation (over 75%)
• Major feedstock for the chemical industry
	• The burning of coal results in pollution:
	  (i) Soot and dust
	  (ii) greenhouse gases
	  (iii) No_x and sulphur compounds which cause acid rain
	  (iv) ash which has to be dumped
	• Coal is a non-renewable energy source
	• Colliery and opencast waste tips are unsightly and can contaminate water resources with acid 'run-off'
	• Every year millions of tons of minestone spoil have to be dumped (mainly at sea)
	• Deep mining can lead to subsidence

Table 1.14 Advantages and disadvantages of natural gas as an energy source

Advantages	Disadvantages
• Readily available to the UK in the short and medium term • Although more expensive than coal, it is more controllable and can be burned more efficiently which more than offsets the cost difference • It is cleaner to use and burn than coal and oil • There is no risk of pollution from spillages • Can be used to generate electricity • Provides feedstocks for the chemical industry • The North sea gas fields create economic wealth on a national basis	• Reserves are substantially less than for coal and oil • Like the other fossil fuels, the burning of gas reduces air quality, produces acid rain and 'greenhouse' gases • Natural gas is a non-renewable energy source • The extraction of on-shore gas reserves can lead to subsidence

Nuclear energy

Two decades ago this was looked upon as the answer to all energy problems. Electricity could be generated on commercial scales comparable with fossil fuel power stations. In the case of fast breeder reactors it appeared that nuclear fuel would be self-regenerating and would be virtually infinite. Further, nuclear power stations produced no toxic or noxious flue gases to pollute the environment. It was a clean fuel. However, it has failed, so far, to achieve its initial promise.

• The decommissioning of worn-out and obsolete power stations causes technical and financial problems for which answers are still being sought.
• The safe storage of spent nuclear fuel is also difficult and costly.

Table 1.15 Advantages and disadvantages of oil as an energy source

Advantages	Disadvantages

• Readily available to UK in short and medium term but reserves locally and worldwide are less than for coal
• Provides employment:
  (i) making oilrigs for offshore extraction
  (ii) operating and maintaining oilrigs
  (iii) building and operating refineries
• Creates economic wealth on a national scale
• A major feedstock for the chemical industry, the refinement of crude oil also produces byproducts such as:
  (i) diesel oil (ii) lubricating oils
  (iii) paraffin   (iv) white spirits
  (v) butane   (vi) propane (vii) bitumen
	• Like the other fossil fuels the burning of oil reduces air quality, produces acid rain and 'greenhouse' gases
	• Oil is a non-renewable energy source
	• Spillages cause environmental pollution killing coastal marine and bird life. Spillages also contaminate the beaches and damage the tourist industry
	• Oil refineries cause visual pollution due to their size and structural requirements

• Public concern over the disastrous accident at Chernobyl in Russia and minor accidents and leaks at many other nuclear installations has led to a powerful political lobby opposing the use of this form of energy in the UK. It is still widely used abroad.

Table 1.16 summarizes the advantages and disadvantages of nuclear energy.

Table 1.16 Advantages and disadvantages of nuclear energy sources

Advantages	Disadvantages
• Clean in use • Produces no greenhouse gases • Does not reduce air quality • No pollution resulting from extraction as is the case for Coal and oil • Creates wealth on a local and national scale	• High initial and operating coats compared with fossil fuel power stations • Problems of disposing of radioactive waste • Problems of decommissioning obsolete nuclear power plants • Radioactive emissions from nuclear power sources (minor accidents) • Radioactive contamination from major accidents such as Chernobyl in Russia

Infinite (renewable) energy sources

For all practical purposes these are inexhaustible energy sources that are self-regenerating such as wind and tidal power or can be regrown such as crops. Another advantage is that many of these energy sources do not cause pollution. Let's now consider the main infinite (renewable) energy sources.

Hydroelectricity (water power)

Water wheels have been used since ancient times to power mills for grinding corn and, in the early days of the industrial revolution, to drive spinning and weaving machines. This was not a reliable source of energy since the machines came to a standstill in the summer under drought conditions and manufacturers quickly turned to steam power when it became available. The water turbine used to drive alternators to generate electricity is the modern equivalent of the water wheel. However, it is much more powerful and efficient.

Lack of suitable water supplies in the UK has limited such an energy source to some regions of Scotland. However, hydroelectricity is used widely in mountainous countries such as Switzerland, and the Alpine regions of France, Italy and Austria. It is also used in parts of the USA, Canada and Africa. An interesting variant is found in Wales where off-peak electricity, generated in conventional fossil fuel power stations, is used to pump water from a low level reservoir to a high level reservoir. At times of peak demand the process is reversed and the water flows back to the low level reservoir through the turbines to generate hydroelectricity.

Hydroelectricity is the only large-scale renewable energy source in use commercially. Other sources are being developed but, as yet, are only in the experimental stage or are operating commercially on a useful but very small scale compared with conventional generation.

Tidal energy

Tidal energy sources makes use of the fact that the tides rise and fall with unfailing regularity twice in 24 hours. The current thinking to harness this form of energy is to build an dam or 'barrage' across a tidal estuary. Reversible water turbines driving generators are built into this barrage. As the tidal waters rise, the turbines are driven by the incoming flood of water from the sea. This forms a huge reservoir of water behind the barrage which is supplemented by the waters coming down the river(s) feeding into the estuary. At low tide the flow reverses as this reservoir of water floods back out through the barrage into the sea. Hence the need for reversible turbines so that they are driven no matter whether the water is flooding in or flooding out. A number of such plants have been built abroad but so far the cost makes them uneconomical compared with conventional methods of generation. Also such installations tend to destroy important wildlife habitats. However, as supplies of fossil fuels run out and the need to reduce man-made pollution becomes more urgent, such schemes become more attractive.

Wave energy

At first sight the use of the waves as they rise and fall to power some form of electricity generating plant seems simpler and less costly than tidal barrages. Further, they do not have to be sited at centres of wildlife conservation (estuaries). However, in practice, the destructive energy of the waves during the frequent storms that blow up round the coasts of the UK have produced engineering problems that have yet to be overcome. Again, the cost involved for the amount of electricity generated outweighs the advantages and at present such schemes are not an economical proposition. Experiments continue.

Solar energy

This is used widely for powering artificial satellites where the sunlight is uninterrupted. Very little electricity is generated by solar energy terrestrially. Generation can be either by direct conversion using solar cells or by using mirrors to concentrate the sun's rays onto heat exchangers to generate steam to drive conventional generating equipment. Unfortunately such plants cannot operate at night when electricity is most needed and, in any case, UK climatic conditions make the availability of adequate sunlight uncertain. At present solar energy is mostly used for heating buildings and water on a local basis.

Wind energy

Wind farms have grown up rapidly in recent years all over the world. The wind turbines (similar to large aircraft propellers) are used to drive generators. Such wind farms cover large areas of ground and cause visual and noise pollution. Such pollution is a nuisance rather than damaging, as is the case when fossil fuels are burnt. However, it may be felt that visual pollution and noise pollution diminish the quality of life for people living nearby. The amount of energy generated in this way is small compared with the amount generated by conventional power stations and small for the amount of room taken up.

Geothermal energy

The temperature of the Earth's core is about 4000°C. By the time this heat energy at the Earth's core has been conducted to the surface through the various rock layers it has cooled significantly. However, evidence of this natural heat energy can be found in:

- hot springs (geysers);
- hot rocks in various locations deep underground (notably Cornwall in the UK).

Experiments are going ahead to tap these sources of energy. Water can be pumped down boreholes to the hot rocks where it flashes into steam which is forced back to the surface under its own pressure to power turbines which, in turn, drive electricity generators. Very little electricity is generated in this way at present. In some countries hot springs are used for district heating schemes but not on a large scale.

Biomass energy

This is the production of heat energy resulting from the burning of *biological* fuels–for example, the burning of wood waste and other vegetable waste or the burning of gas produced by rotting vegetable waste (e.g. landfill gas (LFG)). Providing reforestation and the planting of crops continue at a rate equal to or greater than the rate of use, then this is a infinite (renewable) source of energy. Experiments are also being carried out into the growing of crops for direct conversion to alcohol type spirits for use as an alternative to petrol in motor vehicles in some parts of the world. Although cleaner than petrol and a renewable energy source, all biomass fuels cause some pollution and give off greenhouse gases when burnt.

Table 1.17 summarizes the advantages and disadvantages of the infinite energy sources discussed above.

1.4.5 Energy policies

To preserve our dwindling reserves of fossil fuels until more environmentally friendly and renewable alternatives can be developed, it is essential to have a well-planned and correctly managed *energy policy*. Further, by using existing fuel stocks more efficiently, less pollution is caused and less global warming takes place. We can look at the policies of an overall *energy strategy* under a number of different headings.

Physical energy savings

These are the things which can be done in the home and at work to save wasting fuel. They include installing:

- double glazing;
- cavity wall and loft installation;
- efficient heating controls including thermostats on individual radiators and programmable timers;
- modern boilers that burn the fuel and transfer the heat energy to the water more efficiently;
- automatic doors and draught excluders;
- pipe lagging;
- tank lagging.

Energy audit and monitoring

It is no use installing the energy saving devices listed above if they are not used correctly. Energy audits are used to control the total energy costs of manufacturing by identifying energy waste and should include:

- energy used for heating the premises;
- energy used for process heating;
- energy used for driving machinery.

As with the annual financial audit, it is preferable for the energy audit to be set up by an external professional organization that has no vested interests in the company concerned. The energy audit is set up in conjunction with the energy manager and the chief accountant to analyse ways of:

- identifying energy waste (e.g. unnecessary idle running of machines, heat loss from premises, etc.);
- examining the company's product range to see if design changes can reduce the amount of process energy required;
- planning a strategy for the more efficient use of energy;
- setting improvement targets.

Table 1.17 Advantages and disadvantages of infinite (renewable) energy sources

Advantages	Disadvantages
Hydroelectricity • Clean, no atmospheric pollution • Cheap and constant supply	• High capital cost constructing dams • Ecological damage to flooded areas • often remote from the users • Only 2% hydroelectricity produced in UK for geographical reasons
Tidal Energy • Could be a constant source of electricity • Clean, no atmospheric pollution	• High capital cost makes such generation uneconomical at present • Ecological damage to wetlands surrounding estuaries across which barrages would have to be built
Wave Energy • Clean, no atmospheric pollution • Could be a constant source of electricity	• Experimental work in progress but many technical problems to be overcome
Solar Energy • Clean, no atmospheric pollution • Silent in operation (see wind power) • Safe in operation • Mainly used for water heating in buildings at present	• Costs make it uncompetitive with conventional generation at present • Large areas of land required • Generation zero at night when demand is at a maximum • Sunlight insufficiently constant in UK.
Wind energy • Clean, no atmospheric pollution • Wind is strongest in winter when demand is greatest • Technology relatively simple	• Large wind turbines are noisy and have a limited output • Large areas of land are required for the many turbines required for wind 'farms' • Visual pollution–many large turbines clustered on hill tops are unsightly • Rotating blades can cause local disruption to TV reception.
Geothermal energy • Cleaner than convention energy sources • Some hot springs and geysers are used for district heating schemes but not in the UK	• No suitable sources in UK except possibly in Cornwall where hotrock formations exist deep underground • Long-term effects of removing heat energy from the core of the earth are at present unknown • Generation of electricity by geothermal energy only in the experimental stage at present

(continued overleaf)

Table 1.17 *(continued)*

Advantages	Disadvantages
Biomass Energy (i) Biogas (landfill gas) LFG • Contains up to 60% methane and has a similar calorific value to propane, butane and natural gas • Produced when vegetable matter rots. The residual compost is a useful fertilizer (ii) Burning waste vegetable matter • Waste vegetable matter and waste forestry products are cheap and plentiful • Burning reduces volume of matter to be disposed of	• Atmospheric pollution occurs whenever burning takes place • LFGs are dangerous if allowed to accumulate–should be flamed off or used for electricity generation • Atmosphere pollution occurs whenever burning takes place • Soil erosion and climatic charges occur it proper woodland management and a policy of reforestation is not carried out

A well-organized and successful energy audit should not only pay for itself but should show significant savings that will enhance the profitability of the company as well as reducing the environmental damage caused by the company's operations.

Energy awareness

In addition to company policy leading to the efficient use of energy through the measures listed above and supported by the investment to implement them, a *programme of education* to make the staff aware of the part they can play is also required, that is, the introduction of an *energy awareness* campaign. Simple actions such as turning off unwanted lights, reporting leaks from radiators and hot water taps, closing doors and windows can lead to significant savings. Bonuses paid for useful suggestions leading to energy saving changes in working practices can also help to raise the level of energy awareness.

Energy efficiency

Technological developments have also helped to improve the efficient use of energy, thus reducing the amount of fuel used and the amount of pollution caused. For example:

• Gas fired power stations are environmentally cleaner and more efficient than coal fired power stations.
• Generators powered by gas turbines can be started and stopped as required to augment the conventional power stations at times of peak demand.
• Composite stations use the waste heat from the gas turbine exhaust to generate steam to drive conventional steam powered generating sets.
• The waste heat from conventional power stations can be used to heat nearby factories and homes. Such schemes can increase the energy efficiency of a power station from about 35% to about 75%.
• More attention is being paid to the reuse and recycling of waste materials from production processes and recycling plants are being set up. However, this is more concerned with conserving natural resources

since the energy required for recycling is often as great as that used to produce the raw material in the first place.

- New machinery and plant is increasingly designed to be more energy efficient and more environmentally friendly than the equipment being replaced. Investment in new equipment is to be encouraged.
- The materials used in building modern factories and warehouses provide better heat insulation. They also require less maintenance and less energy is used in their manufacture in the first place.
- Modern fluorescent lighting is five times more efficient than conventional filament lamps as well as giving more even illumination.

Test your knowledge 1.21

1. Explain the main differences between finite and infinite energy sources and give an example of each.

2. Briefly report on the main pollutants associated with the burning of fossil fuels and the damage such pollutants can do to the environment.

3. Briefly report on the advantages and disadvantages of using hydroelectricity, tidal energy, solar energy and nuclear energy for the generation of electricity compared with conventional fossil fuel burning power stations.

4. Briefly report on the measures being taken in your school or college to conserve energy and, if possible, suggest how current practices could be improved.

Key Notes 1.4

- From the days of the most primitive communities, manufacturing has always adversely affected the environment.
- Manufacturing impacts on the environment in a number of ways by affecting our: quality of life; natural resources; renewability of resources; local and national geography; pollution; and local and national economies.
- Pollutants can be: *airborne* (particles, toxic gases, greenhouse gases, odours and noise); *waterborne* (deliberate and accidental discharge of chemical and untreated sewage into lakes, rivers and estuaries, and the leaching of the animal wastes and nitrate fertilizers used in agriculture into rivers and streams); *visual* (spoil heaps and pithead gear of coal mines, derelict industrial buildings, disused quarries, the insensitive erection of large industrial plants in areas of natural beauty).
- *Greenhouse gases*, such as carbon dioxide, collect in the upper atmosphere and form a 'blanket' that upsets the balance between the heat energy received from the sun and the heat energy that can escape into space. This 'greenhouse effect' causes the surface temperature of the Earth to increase and can lead to flooding if the polar ice caps start to melt. It also causes climatic changes on a global scale.
- Destruction of the *ozone layer* is often confused with global warming; however, it is a totally different phenomenon. Destruction of the ozone layer is caused by the CF gases originally used in refrigerators and aerosols. The ozone layer limits the intensity of the ultraviolet radiations from the sun reaching the Earth. In excess, these radiations are dangerous and damaging to all living creatures including ourselves.
- Unlike domestic waste, *industrial waste* is classified as *controlled waste*. This can be divided into *hazardous waste* (e.g. toxic, narcotic and flammable substances) and *nonhazardous* waste (e.g. ash and builders' rubble).
- *Reusable waste* can form the raw materials of other industries–for example, power station ash can be made into building blocks.
- *Recyclable waste* can be reprocessed and reused–for example, scrap metal can be melted down and reused. Waste paper can be pulped and turned back into reusable paper and board products.

- *Degradable waste* is broken down by bacterial attack or by strong sunlight. Vegetable matter is broken down by bacterial attack to make horticultural composts. Some plastics are designed to be broken down into a harmless ash by the effects of strong sunlight.
- *Byproducts* are secondary products of the production process. They are saleable products in their own right–for example, when crude oil is refined to make petrol, lubricating oils and diesel oil (derv) are recovered as byproducts. Their sale helps to keep down the price of the main product (in this case petrol).
- In the UK the most common means of waste disposal are: landfill sites (solids); incineration (much domestic and light commercial waste); disposal at sea (non-toxic and solid and liquid waste); fresh water courses and estuaries (suitably treated liquid effluents); and composting (vegetable matter).
- Landfill sites must be properly engineered so that they are safe, stable and suitable as agricultural, recreational or building land when no longer required.
- The main pollutants associated with landfill sites are: landfill gas (LFG) which is created by rotting vegetable matter and is highly flammable; leachate (noxious liquids) which must be contained within the site and not allowed to pollute the surrounding ground and water courses.
- Finite energy sources cannot be renewed once they have been used. This applies to all fossil fuels such as oil, coal and natural gas.
- Infinite (renewable) energy sources are not used up if properly managed–for example: hydroelectricity; tidal energy; wind energy; wave energy; solar energy; and biomass energy sources.
- *Acid rain* is produced when gaseous oxides of nitrogen and sulphur, produced during the combustion of fossil fuels, react with water vapour in the upper atmosphere.
- *Nuclear energy* is derived from controlled atomic reactions. These reactions produce heat energy which can be used to generate steam for the conventional generation of electricity. Theoretically this is the cleanest method of generating electricity that is economically comparable with conventional power stations. Unfortunately the accident at Chernobyl in Russia and a near disaster at Three Mile Island in the USA, coupled with the difficulties and cost of safely disposing of nuclear waste and decommissioning worn-out reactors, has made this form of electricity production unacceptable in the UK.
- Companies need to have well-managed *energy policies* to implement their *energy strategy*. This is designed to save dwindling fuel resources, reduce global warming and pollution. Indirectly it saves the company money.
- An *energy audit* is required to ensure that the energy strategy is being correctly managed and successfully implemented. Like a financial audit it should be carried out by an independent firm of specialists.
- *Energy awareness* is achieved through staff training and education. This is essential if the energy strategy is to be properly and effectively implemented.
- *Energy usage efficiency* is achieved through investment in, and the correct use of, plant that reflects the latest technological developments. The savings brought about by the use of such plant can recover the cost of the initial outlay.

Evidence indicator activity 1.4

1. Write a report on the environmental impact of the production of two contrasting products with different scales of production. Your report should cover:

   - the impact on the environment (human, natural and built);
   - the impact on the quality of life, local economy, renewability of resources, local geography and pollution;
   - any waste that is generated and its disposal;
   - the energy sources used and the identification of the benefits of an organizational energy policy as applied to a given manufacturing process.

Chapter 2

Working with a design brief

Summary

This chapter is covers the requirements of Unit 2 of the GNVQ Manufacturing (intermediate). The *unit* is subdivided into two *elements*. The first element (2.1) is concerned with the creation of *product proposals* from a given *design brief*. The second element (2.2) is concerned with the finalizing of design proposals using feedback from presentations.

2.1 Originating product proposals from a given design brief

2.1.1 Introduction

Today we live in a society that is dependent upon technology. Our standard of living and the way we live rely heavily upon the use of products, designed and manufactured to make our daily lives easier. All the articles around us have been designed by someone attempting to solve a particular problem – for example, the next time you get up and prepare to go to school, college or work, think about the products around as you wash, dress and have breakfast before leaving home. The clothes you will be wearing, the furniture and appliances in your kitchen, the cooking utensils and food you need for your breakfast, as well as the television and radio you may turn on as you get ready to go out, all have been designed and manufactured in response to a particular need.

Nowadays there is a wide range of materials and information available to us, together with the knowledge of many technologies and skills to help us when designing new products. We also have automated equipment to produce our products in great quantities, more rapidly and to the required quality standards.

We have already seen in the previous chapter that the technologies available to us can be harmful as well as beneficial, for example power generating technology allows us to produce electricity in great quantities, but the burning of fossil fuels contributes to producing acid rain. Increasingly, product designers will have to think about the potentially harmful impact of their products on the environment we live in.

The role of design

The key role of design was highlighted in 1988 by *The Engineering Council* in its booklet, *Managing Design for Competitive Advantage*, where it states that:

A product sells when quality, price and availability match customer requirements. As international competition intensifies, the challenge of new products offering improved value increases.

Figure 2.1 highlights the key role that design plays, whether the company is engaged in creating mass-produced products, chemical engineering processes, communication systems or one-off civil engineering projects.

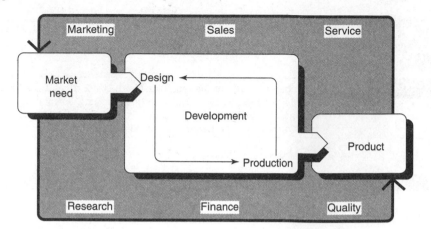

Figure 2.1 The key role of design.

Design is the process of preparing the detailed information which defines the product to be made. Production is the process of using the resources of materials, energy and manufacturing skills to convert that information into physical components which, along with purchased items are then assembled into products or systems.

Development is the process of acquiring specific information about the performance of a proposed product by constructing and testing prototypes. This information can be used to redesign the product and improve the manufacturing methods before starting full-scale production. The process of development is therefore inseparable from the processes of design and production.

Fundamental to all commercial organizations is the effective management of personnel and finance. The engineering activities of design, development and production are also supported by, and should have close links with, marketing, research, quality, sales and service – the relative importance of each depending on the particular market and the individual company.

The responsibility of the design team should not end when the production documents are issued. There should be a *continual feedback of information* from the production, sales, service and marketing teams, in order to build up as much knowledge as possible about the product and its performance in the market. This allows timely improvements to be introduced in order to stay ahead of the competition, and also stimulates ideas for new products.

In this chapter we will see that design is about:

- identifying needs and opportunities;
- generating design ideas and solutions;
- utilizing our knowledge of technologies, materials and manufacturing processes in order to realize the design;
- satisfying customer demands (requirements).

2.1.2 Development of a design brief

In chapter one we defined the key purposes of manufacturing *as creating wealth* and *satisfying consumer demand*. The design process starts as soon as a consumer (customer) need is identified. The consumer need is translated into a product idea which is followed by a project proposal and a feasibility study. The feasibility study is used to develop the *design brief*.

Working from the design brief, production constraints and quality standards are identified. A number of design proposals are then developed. Each design proposal is assessed for feasibility.

The preferred design proposals are presented to the client for approval. Client feedback is used to prepare the final design solution. The stages that lie between the customer's requirements and a realistic design proposal are shown in Fig. 2.2.

Let's now look at these stages in more detail.

Identifying consumer need

The product must meet the needs (requirements) of the customer as closely as possible. This is often a compromise between the ideals of the customer and what can be realized within the technology available and the price the customer is prepared to pay. For a ship, an oil rig, or a bridge for instance, the customer has expert knowledge and will produce a detailed specification to set the designer on his or her way. On the other hand, products such as cars and domestic appliances have to satisfy the requirements of very many people all of whom will have their own idea of what they want from such a product and how much they are prepared to pay. Market researchers will carry out surveys of what is available and what features need to be changed to attract customers to the new product under consideration. Many of the requirements will conflict with each other and the final design will have to be a compromise.

Consumer need for a new product may be triggered by any of the following:

- A specific enquiry from a customer.
- The development of a new technology.
- A new method of applying existing technology.
- Identification of a gap in the market.
- A decline in the sales revenue generated by existing products resulting from a reduction in demand.

Next the consumer need is translated into a brief description outlining what the product has to do and what its commercial potential may be. In many industries this is called the *product concept*.

If the product idea or concept is thought to have commercial potential, a *feasibility study* usually follows. The feasibility study should contain sufficient information to allow senior management to approve or reject the product idea.

Design brief

This is a description of the client's requirements given to the designer. It should list the requirements in terms of product performance, that is,

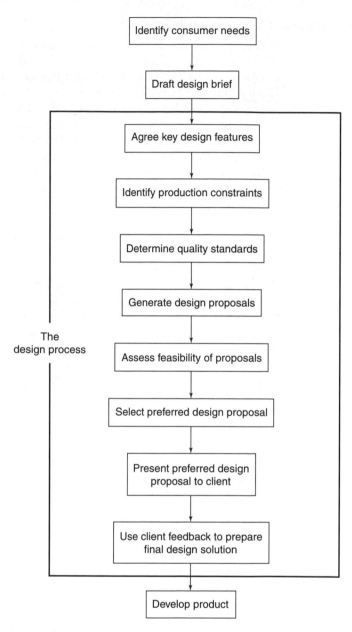

Figure 2.2 The design process.

functional details, costs and any special requirements such as times scales. It will also list quality standards and safety regulations that the product designer must observe, but must not provide design solutions.

A design brief may describe the updating of an existing product or the need to redesign a product to reduce production costs, or a product to be designed from a completely new concept. The key requirements of a design brief can be summarized as:

- the customer's concept;
- the functional requirements;
- market details such as affordable cost, market size, safety legislation and legal requirements;
- quality standards.

Once the design brief has been agreed, the work of designing the product can begin.

Test your knowledge 2.1

1. Draw up a design brief for a simple, low cost product of your own choice – for example, a bird table.

2. Draw up a design brief for improving some article with which you are familiar.

2.1.3 Key design features

Having established our design brief, the designer now has to set to work on an initial concept. Let's now look at some of the factors the designer needs to consider when developing the design brief into a working design. These factors can be summarized as:

- aesthetic;
- contextual;
- performance;
- production parameters;
- quality standards.

Aesthetic

Aesthetic features are concerned with the appreciation of the beauty of products. In other words, features that make a product look good. Ideas about what looks good will vary from person to person. But the judgment of what makes a product attractive to look at or pleasing to use or consume is based on the senses of sight, touch, taste, smell and hearing.

- Sight is influenced by the look of a product, that is, its shape, form or colour.
- Touch is influenced by the feel of a product, that is, its texture or surface finish. Hand-held products need to be designed to fit the shape and dimensions of the human hand.
- Taste can be either sweet, sour, bitter or salty. Smells can be pleasant and exciting or unpleasant and annoying. Smell is usually related to taste.
- Hearing can be irritated by squeaks or rattles, with loud noises becoming intolerable. Excess noises can be damaging to health. In the case of entertainment equipment, such as CD players, the quality of the sound is very important.

Having established that *aesthetic* features are those that make the design attractive to look at and pleasing to use, by way of an example, let's compare a Victorian mangle you may have seen in a museum with a modern wringer. The mangle would be made from wood and cast iron. It would be painted black and would be heavy, bulky and very strong. To our eyes it would be ugly and 'overengineered'. It would have unguarded gears and would be clumsy to use. Aesthetics would not have been considered in its design. It would simply be functional.

However, the modern wringer will have been designed to satisfy a more discerning market and will be judged in competition with other modern household appliances. Aesthetics will most certainly have been considered in its design. Its lines will be smooth and pleasing to the eye. It will be easy to keep clean and light and easy to use. Modern materials will have been used and there will be no dangerously exposed moving parts. It will be light in colour and small in size to fit into a modern kitchen. Nevertheless it will still be functional.

We can summarize our present aesthetic design ideas as being based upon:

- simplicity of shape;
- smooth lines;
- balance and proportion;
- attractive colours;
- the ability for a product to blend in with its environment;
- how pleasant they are to our senses (hearing sight smell, taste, touch).

Contextual

Contextual design features are those features that have to be included to meet the demands of the general setting, situation or environment in which the product is to be used, that is, the relevance of a design to its working environment and the persons who will be using it. Some of the contextual factors that must be considered could be:

- Will the product be used indoors or out of doors?
- If indoors, will it be used in a damp (kitchen or bathroom) or dry (living room) environment?
- Will it be used in a temperate or a tropical climate?
- What is the target age group?
- Will it be used by persons suffering from disabilities?
- Will it be aimed at a specific market (male or female) or will it have universal appeal?

Performance

The performance of a product is measured by how well it functions when used – for example, the performance of engineering products is frequently defined by criteria such as:

- input voltage or current;
- mechanical forces applied;
- operating life;
- operating speed range;
- operating temperature range;
- output power range;
- efficiency;

- performance accuracy;
- running time between services.

On the other hand, the performance of clothing products is more likely to be judged on their ability to:

- resist wear and tear;
- resist shrinkage when cleaned or washed;
- resist fading;
- resist creasing when worn;
- remain smart in appearance over a reasonable period of time.

Therefore it becomes apparent that the performance criteria varies with the product under consideration.

Production parameters

Production parameters are influenced by factors such as:

- size;
- weight;
- cost;
- quantity;
- time.

Size

The size of a product will influence the cost of materials used. So to keep material costs to a minimum designers will try to keep product size as small as possible, so long as product function is not affected. If size reduction demands more precise manufacturing methods, product cost may increase.

Size will also influence the space required to manufacture the product – for example, aircraft need much more space than microelectronic products. Manufacturing space requirements will influence overhead costs and eventually total manufacturing costs.

Weight

The design weight of a product will also influence its size and the type of material used – for example, to keep weight as low as possible aluminium alloys are used in the manufacture of aircraft.

Cost

As already seen costs are an important factor, as almost any production parameter can be related to cost. The total cost of a newly designed product is the sum of such cost elements as:

- design costs;
- development costs;
- production material costs;
- production manufacturing costs.

The designer needs to know the magnitude of these costs so that he or she can make informed choices between different materials and different production processes. The probable demand is also important since this will also influence the production process chosen.

Quantity

The scale of production used, that is, single item, small batch, repetitive batch or continuous flow, will be influenced by the demand (product quantity). As we saw in Chapter 1, this had a significant effect on the manufacturing processes chosen and on the cost of manufacture.

Time

This is related to the production process and influences the cost of the product. The greater the manufacturing time per unit of output, the greater will be the cost of the product. Conversely, the shorter the manufacturing time, the less will be the cost.

Quality standards

The client specifies product quantity standards. To assure that the product *conforms* to the specified quality standards, careful attention must be paid to their selection by the client. Some typical quality standards are listed below.

- Product performance standards.
- Material specifications.
- Manufacturing tolerances.
- Surface finish requirements.

Product quality effects cost. Generally, a low quality specification can be equated with low cost. Conversely a high quality specification can be equated with a high cost. Ideally, the product designer produces a design that strikes a balance between these extremes, so that the product has adequate quality and will conform to the requirements of the consumer but, at the same time, is suitable for manufacture at a cost that the customer can afford.

Test your knowledge 2.2

1. Choose an article of clothing or a household appliance with which you are familiar and briefly assess it in terms of:

   (a) aesthetic design;
   (b) contextual design;
   (c) production parameters.

2.1.4 Production constraints

Once key design features have been agreed between client and designer the next area to consider are the *production constraints*, that is, any factor within the following areas that may make it difficult or impossible to manufacture the newly designed product. These areas include:

- available labour;
- available materials;

- available technology;
- health and safety requirements;
- quality standards of production equipment.

Available labour

A newly designed product may require large numbers of production personnel with skills not presently available within the existing workforce. To overcome this problem manufacturing management may decide to recruit more workers with the required skills or retrain (reskill) existing workers. Alternatively, the designer may have to modify the proposed design, so that it can be produced by the existing workforce. If more workers cannot be employed and a design change is not feasible, then the new work may need to be subcontracted out. In some instances, where a product is being redesigned to take advantage of new technological developments or a change in customer requirements, the workforce may need to be reduced.

Available materials

The new product should be designed to use readily available materials (standard sizes and specifications) which can be purchased easily and quickly from a material stock holder. Readily available materials should keep material costs to a minimum. Materials that are difficult to obtain are likely to be more expensive and may hold up manufacturing if they are not delivered when required.

The designer's choice of materials will influence the manufacturing process that is eventually chosen to produce the new product. Remember, very hard and very soft materials are sometimes difficult to shape by machining processes. The former cause excessive tool wear and the latter tend to tear and drag leaving a poor surface finish.

Modern manufacturing tends to employ a near finished size (NFS) philosophy so that machining processes are kept to a minimum – for example, sintered powder metal compacts may be used in place of forgings. Although basically more expensive, this extra cost is more than offset by the lack of need for expensive machining operations. The connecting rods of certain BMW car engines are made in this way. Another example is the use of shell moulding and investment casting in place of traditional green sand moulding to give greater accuracy and a better surface finish. Again this reduces the need for costly machining and finishing processes.

Available technology

To maintain competitive advantage new products should be designed to make as much use as possible of new technology providing it is well proven and appropriate for the product. New technology should not be used just because it happens to be available. It should only be used if it will provide a better quality product and a more economical price.

For example, wooden tennis and squash racket frames have been replaced first by aluminium and then by carbon fibre reinforced plastics to give a stronger and lighter product that is not only manufactured more easily, but performs more efficiently.

Soft drinks and beers/lagers are now frequently supplied in lightweight aluminium cans instead of heavy glass bottles. Not only are the cans manufactured quickly and cheaply using advanced manufacturing techniques, but considerable savings can be made on transportation costs due to the lighter weight.

If the manufacturer decides to install new manufacturing equipment to produce the new product it is likely that the existing workforce will require familiarization and retraining in new manufacturing processes. This may initially slow down production output if problems are encountered.

Health and safety requirements

The Health and Safety at Work Act of 1974 places the responsibility for safety equally upon:

- manufacturers and suppliers of manufactured products and equipment;
- employers;
- employees.

So manufacturers have to ensure that all products are manufactured within relevant health and safety laws. Any manufacturing processes that can cause damage to the health of employees or to the environment can give rise to a production constraint in the form of a Prohibition Order issued by an inspector of the Health and Safety Executive (HSE) until the process can be made safe or replaced by one that is less hazardous to the satisfaction of the inspector. For example:

- chemical hazards may create burns and toxic fumes that can cause injury;
- electrical hazards can cause burns and shock;
- extreme conditions of heat, humidity or cold can cause stress and fatigue;
- equipment giving off radiation can cause tissue damage;
- environment pollution can be caused by the production of scrap materials, waste products and excessive noise from manufacturing processes;
- exposure to excessive noise can cause damage to employees' hearing.

The product itself must also conform with the appropriate British Standards Institute's (BSI) recommendations for safety in use. It must also conform with the European safety requirements and legislation and carry the appropriate CE markings.

Quality standards of production equipment

The product quality will have been agreed with the customer during the generation of the design brief. The designer will have had to conform with this quality requirement at every stage of the design process. Similarly the production equipment or machines chosen to produce the components to be used in a newly designed product must be capable of consistently meeting the product designer's specified requirements for dimensional accuracy, geometric accuracy and surface finish. In other words, the chosen production equipment must have adequate *process capability* to consistently manufacture the required product.

Test your knowledge 2.3

1. State what is meant by the term 'production constraint'.

2. State who have legal responsibilities under the Health and Safety at Work Act.

3. Find out what constraints are available to an HSE inspector and under what circumstances they are used.

4. Explain briefly under what conditions members of a workforce may require training to reskill them. Suggest why this may be more satisfactory than replacing them with workers who already have the required skills.

2.1.5 Generating proposals to meet the design brief

A design brief calls for a small hand-held, portable electric fan that can direct a gentle stream of cool air onto the face of the user while watching sporting events on hot sunny days. The basic requirements are for the fan to be light in weight, compact, pleasant to hold, reliable, efficient, be powered by a readily available source of energy and be capable of being manufactured and marketed at a relatively low cost.

These requirements are then developed into a formal design brief such as that shown in Table 2.1.

Table 2.1 Design brief (key design features)

Product:	Hand-held, portable electric fan
Aesthetic requirements:	• Comfortable to hold by hand for extended periods of time • bright, primary colours for easy recognition • tough, durable and compact • light in weight • quiet running
Contextual requirements:	• to be used out of doors in hot locations • to be used in temperate and tropical countries • equally attractive to men, women and children • safe for children to use • easy to change batteries
Performance:	• to be powered by dry batteries (2 × 1.5 V type AA) or by solar cells • low current consumption for long life • must provide adequate ventilation • reliable, durable, capable of achieving a long service life
Production parameters:	• low cost – under £12.00 • initial batch size 5000 per month • quality standards – minimum battery life of 2 hours' continuous running – moulding finish to be comfortable to hold and free from sharp edges – to be moulded from a tough, impact resistant plastic

Figure 2.3(a) shows one possible design. The fan could be driven by a small direct current (d.c.) motor which is powered by two size AA dry cells

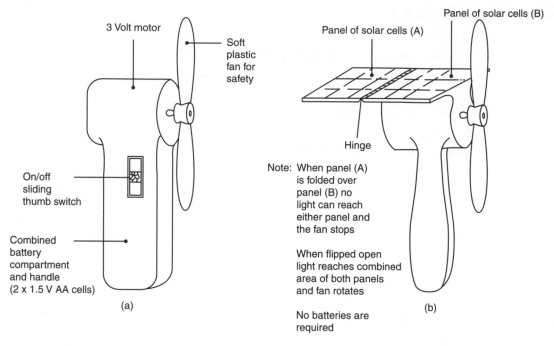

Figure 2.3   Hand-held electric fan.

connected in series to give an output of 3 volts. The fan operates when the device is switched on.

Figure 2.3(b) shows an alternative design. The fan is again driven by an electric motor but the energy source is a solar panel built into the case of the fan. No switch is provided since the fan can be stopped simply by covering the solar panel.

We have now generated two possible designs. In order to assess their feasibility we have to draw up a comparative table of the design proposals similar to Table 2.2.

Test your knowledge 2.4

1. For a simple product of your own choice (it can be quite simple, such as a confectionery item, a garment, or a simple tool):

   (a) sketch the product;
   (b) produce a design brief for the product;
   (c) tabulate the key design features;
   (d) tabulate the process constraints for its manufacture.

2.1.6 Assessing the feasibility of design proposals

The first step is to assess the alternative design proposals such as those outlined in the previous section. The factors to be considered will vary from product to product, but the main factors common to most products are:

- cost;
- marketability;

Table 2.2 Design proposals for a hand-held portable electric fan

Key design features	Battery powered	Solar powered
Size	comfortable to hold and compact if fan blades are made to fold when not in use	More bulky than when battery powered because of the solar cell panels, particularly when open
Weight	Approximately 150 g with batteries	Approximately 120 g
Power source	2 × 1.5 volt AA size dry batteries	2 solar cell panels
Running time	Minimum of 2 hours specified. Actual time in excess of 2 hours depending on type of battery used: standard/long life/alkaline rechargeable	Continuous and indefinite if light level is sufficient to activate the solar cells
Motor	Permanent magnet 3 V, d.c. low consumption motor	Permanent magnet 3 V, d.c. low consumption motor
Ventilation	Constant output from the fan over the life of the batteries	Varies depending on level of illumination. Output generally lower than for batteries
Ease of operation	• Thumb operated slide switch on handle • Batteries inserted into handle by removing 'click-on' base	• No switch required. Fan starts immediately solar panel is flipped open, Fan stops when panel is closed • No batteries to change
Reliability	• Indefinite life. Oil impregnated bearings require no maintenance • Batteries should be removed when not in use for long periods • Proven technology	• Indefinite life. Oil impregnated bearings require no maintenance • Proven technology • No switch contacts to wear out
Durability	• Moulded from tough, high impact resistant plastic such as Acrylonitrile-butadiene-styrene (ABS)	• Moulded from tough, high impact resistant plastic such as Acrylonitrile-butadiene-styrene (ABS)
Process constraints	• No reskilling required – plastic mouldings and motor units will be outsourced • ABS to be used as it is impact resistant • Plastic moulding is an available technology • Health and safety. Fan blades are designed from soft plastic and the rotational speed is limited for safety • Safety clothing will be worn by process workers during manufacture	• No reskilling required – plastic mouldings, motor units and solar panels will be outsourced • ABS to be used as it is impact resistant • Plastic moulding and solar panels are now available technologies • As for battery powered fan • Safety clothing will be worn by process workers during manufacture

- materials;
- production requirements;
- quality standards;
- suitability.

Table 2.3 compares the feasibility assessment for both of our fan designs. Both have their own advantages and limitations and it could be difficult to make a decision.

If the feasibility assessment does not give a clear-cut indication as to which design should be selected, then the next stage is to make prototypes for more

Table 2.3 Feasibility assessment of the design proposals

Assessment	Battery powered	Solar powered
Production	• The plastic casing can be an injection moulding using available technology • Assembly methods need to be considered at the design stage as assembly is labour intensive • Use 'Snap-together' components and adhesives, avoid screwed fastenings wherever possible • Use compression joint connections rather than soldering • It is feasible to produce	• The plastic casing can be an injection moulding using available technology • Assembly methods need to be considered at the design stage as assembly is labour intensive • Use 'Snap-together' components and adhesives, avoid screwed fastenings wherever possible • The hinge connecting the solar panels could be a problem • Use compression joint connections rather than soldering • It is feasible to produce but some further work is needed on the hinge connecting the solar panels
Materials	• The motor and switch are standard components that can be outsourced cheaply from the Far East • ABS high impact plastic will be used for the mouldings • It is feasible to obtain parts and materials	• The motor and solar panels are standard components that can be outsourced from the Far East • ABS high impact plastic will be used for the mouldings • It is feasible to obtain parts and materials
Cost	• It is feasible to manufacture for less than £12	• The feasibility of manufacturing for less than £12 will depend upon being able to source low cost solar panels of suitable quality. Also on simplifying the design of the hinge
Marketability	• There is a market for the battery powered fan in the UK and abroad • Some difficulty in obtaining batteries in Third World countries • Can be used indoors and in poor light – an advantage in the UK • It is feasible to market this produce	• There would be little demand for the solar powered fan in the UK • There could be a demand in sunnier and hotter countries • Would be popular in Third world countries were batteries are scarce and the added bulk of the solar panels would not be a problem • It could be feasible to market abroad if price could be reduced, possibly by licensing manufacture to a Far East low cost country
Quality standards	• The mouldings can be produced to the required finish and toughness • Reliable motor units and switches are available • Battery life exceeds the requirement of the design brief • It is feasible to produce this battery operated fan to the specified quality standards	• The mouldings can be produced to the required finish and toughness • Reliable motor units are available • The solar cells are prone to damage. The design needs to be revised to give them greater protection • Subject to some design modifications it should be possible to meet the specified quality standards
overall assessment	• The design for the battery operated fan is acceptable • The market is available • This product is approved for prototype production	• It was felt that this was an innovative design that required further development • In view of the size of the potential overseas market further design and marketing studies were commissioned

detailed evaluation. This evaluation enables the products to be subjected to a range of tests incorporating the extremes of service conditions the product is liable meet in use. It is also possible for prototypes to be tested not only by the manufacturer and by the customer originating the design brief, but also by members of the public to assess their reactions to the alternative designs. Sometime even the most favoured design has to be modified before it is finally considered fit for production.

Ideally, when producing assemblies for test, it is advisable to use the plant and the equipment that will eventually be used in full-scale production. This way, the design team can evaluate the adequacy and technical feasibility of the product prior to full-scale manufacture. It also enables them to carry out a *process capability study* of the equipment to identify if the allocated machinery is capable of holding selective tolerances and providing the required quality of finish. A process capability study is a practical assessment carried out under controlled conditions to determine whether the machinery allocated is capable of working to and holding the required tolerances.

Selecting preferred designs

Having assessed the design proposals and, possibly, built completed prototype assemblies in order to carry out feasibility tests, it is necessary to select the preferred design. This is done by carrying out a range of *design reviews*.

- Design reviews should be conducted by a team of specialists not normally associated with the development of the design.
- The review team should include specialists such as field engineers, reliability and quality engineers, method study engineers, materials engineers, and representatives from the procurement, marketing, costing and packing and shipping departments.
- The review team should work to clearly defined criteria including customer requirements, industry standards, national and international regulations.
- The review team will evaluate areas of reliability, performance, ease of maintenance, interchangeability, installation, safety, appearance, cost, value to the customer and ergonomics. Ergonomics is defined as 'the study of economics in the human environment'. Put more simply, is the product convenient to use? For example, the flight deck of an aircraft or the driving position of a modern car is *ergonomically designed* so that the controls fall readily to hand and are easy to use with the minimum of movement and effort.

Test your knowledge 2.5

1. You are a member of the design review team assessing our two prototype hand-held fans. Write a brief report on your findings and make a recommendation for manufacture giving the reasons for your choice.

Key Notes 2.1

- Modern society depends increasingly on technology.
- The products of technology need to be designed to suit the needs of the consumer.

- Design is the process of preparing detailed information that defines the product to be made.
- Development is the process of acquiring specific information about the proposed product by constructing and testing prototypes.
- New designs are the result of identifying a consumer need that is not currently satisfied.
- The *design brief* summarizes the requirements of the customer's concept, its functional requirements, market parameters such as cost and quantity, and quality standards.
- The design brief is the starting point for the product designer.
- Design aesthetics are concerned with those things that make a product attractive to look at and pleasing to use.
- Contextual design features are concerned with the general setting, situation or environment in which the product will be used and the consumer groups that will use it.
- Performance is concerned with how well the product functions when used. This will vary from product to product.
- Production parameters are concerned with such physical factors as weight, size, cost, quantity, rate of output.
- Quality standards are concerned with such factors as performance standards, material specifications, manufacturing tolerances, surface finish requirements. This can be summed up as *fitness for purpose*.
- Production constraints refer to the availability of labour, materials, manufacturing technology, health and safety requirements and quality standards of the production equipment.
- Health and safety laws apply not only to the well-being of the persons involved in manufacture but also to the consumer and the environment as a whole. Responsibility rests equally on the employer, the employee and the suppliers of equipment and materials.
- Design proposals are developed from the design brief.
- The design proposals are assessed for feasibility in order to select the design that will eventually be manufactured.

Evidence indicator activity 2.1

Develop alternative design proposals suitable for contrasting scales of production for an original product of your own choice in detail and one product in outline. The record of your proposals should include:

- a summary of the key design features, production constraints and quality standards, including consideration of all the key design features for each product;
- at least two proposals for each product, showing how they meet the requirements of the design brief, with details of their key design features;
- a summary of the assessment of these proposals using all the feasibility criteria in the range of this element;
- notes on how these assessments are used to agree the final product selection.

2.2 Finalizing proposals using feedback from presentations

2.2.1 Support material for presentations

At various stages during the design process it is necessary for individual designers to present their ideas to the rest of the team and for the team leader to present their ideas to the customer. For such presentations to be successful they must be carried out in a planned and professional manner. Invariably they will require the use of various audio-visual techniques.

Reports

Reports are essential forms of communication in business today:

- *Individual reports* are used for internal routine communication of day-to-day issues and are written in the *first person*. (I would be grateful if you will attend, etc....)
- *Formal reports* are used to communicate with such persons and organizations as your superiors and customers. They are written in the *third person*. (The sales figures for this month indicate that ... it is unfortunate little improvement is shown, etc....)
- *Legal reports* such as accident reports, are often completed on preprinted forms which only require the respondent to fill in the sections indicated in response to a series of formal prompts. The style of writing (first person or third person) will depend upon and reflect the nature of the prompts and questions.
- *General reports* are usually written in the third person and are commissioned by an official body or person. Examples are shareholders' reports from the chief executive of the company, design reports to the technical director or a customer, student reports to his tutor.

Whichever type of report is required, it should be written to a basic set of rules to ensure that it is logically constructed in the accepted format. For a lengthy report, it should be divided into numbered sections which should be prefaced by a *list of contents*. The items in the list of contents should be numbered to agree with the section numbers. An *index* of key points should be included at the end of such a report. The basic format and order of presentation for a report is as follows:

- the title which must convey the contents of the report (e.g. Design for an electrically propelled saloon motor car);
- the person(s) or body commissioning the report and to whom it should be addressed. In reports for public distribution the address is normally omitted;
- if the report refers to a meeting (e.g. minutes of a meeting) a list of the persons attending together with their titles should be included. A list of persons apologizing for non-attendance should also be included;
- a brief reference to the terms of reference under which the report written. For short internal reports to a single specific person this can often be omitted;
- a summary of what the report contains and its conclusions should be inserted near the front as an overview of the entire report;
- the main body of the report detailing the investigations carried out, the facts discovered, assumptions, arguments and opinions arising from the investigations;
- the conclusions of the writer set out in a clear and logical manner. These should be kept as brief as possible;
- the recommendations of the writer that need to be carried out in order to meet the needs of the report;

- an appendix to the report which contains charts, statistics, graphs and any other information that would be a problem if included in the main body of the report. Such material should be numbered so that it can be referred to in the text of the report;
- for a technical report a glossary of the terms used may be necessary if the intended readership is broad and contains persons with a non-technical background;
- finally the names and signatures of the authors of the report and the date of its preparation should be added.

Notes

Nowadays persons are often called upon to give *presentations* to senior management, customers, financial sponsors, the media and other interested parties. The more professional the presentation, the more influence it will have and the more successful it will be. Therefore, when making a presentation, have notes available and leave nothing to chance. However, remember that notes are for personal use only. They are only an *aide-mémoire* for the speaker. Don't just stand and read your presentation from them.

- Write out your presentation in full – not only as a starting point to make sure you have covered all the important information in the correct order but as a useful prop if your nerves fail you and your mind goes blank!
- Prepare 'prompt cards' from your detailed presentation. Only one topic area on each card with the heading in bold type.
- The prompts cards should also remind you which visual aid you intend to use and the point where you intend to use it. It is useful for these to be colour coded.
- The prompt cards should also carry brief and clear notes summarizing your full write-up with the points highlighted.

Visual aids

At one time or another we have all suffered the 'lecture' where some erudite visiting speaker stands up and drones on and on reading from his or her notes in a monotonous voice, while the chair we are sitting on seems to get harder and harder.

We should always try to liven up our presentation with various visual aids. This is not only to relieve the monotony by introducing a new interest from time to time but also because 'one picture is worth a thousand words'. A picture of an elephant is much more informative than a written description. There are various types of visual aid available to us and first we will consider the overhead projector (OHP) transparency.

Overhead projector transparencies

For a quick, one-off presentation for your colleagues OHP transparencies can be produced on acetate sheets using special transparent colour pens. However, for more formal presentations something better is required. 'Letraset' dry transfer lettering looks a lot more professional but is only available in black. However, such a transparency can be livened up by using coloured acetate for the background. For the most professional presentation computer graphs can be pressed into service using a colour printer. For the experienced computer 'buff' the possibilities are endless using readily available modern software.

OHP transparencies are best used for headings, tables, charts, graphs and line diagrams. For the best results the house lights should be dimmed. This makes it difficult for note taking, so printed copies of all the transparencies should be made available for your audience. They should be made aware of this facility at the start of your presentation. There is nothing more annoying than to have struggled to take notes in the dark only to be informed of the availability of handouts at the end of the proceedings!

Visual aids (35 mm transparencies)

Colour transparencies made on 35 mm camera film are the best way of showing pictorial material such as machines, cars, furniture, clothes, etc. These can be copied from previously published sources (with the permission of the copyright holder) or from the actual objects shown. Remember to check that they are correctly inserted in the projector magazine prior to the presentation. Nothing looks more slipshod than slides that are upside down, reversed left to right or in the wrong order. It makes a very bad impression on your audience no matter how much care you have taken over the rest of your presentation.

Visual aids (video recordings)

With the advent of high quality portable camcorders video recordings have virtually superseded ciné-film where moving pictures are required, particularly where sound is required as well. Professionally produced they are widely used for advertising and for staff training and reskilling. Compared with ciné-film, video recordings have the following advantages:

- They are quick and relatively cheap to produce and copy.
- The sound is easy to redub so as to produce various foreign language versions to suit particular overseas markets.
- Before the advent of video recordings, a company designing and manufacturing large plant and heavy machines had to bring its overseas customers to this country to see their products or transport expensive and heavy working models to the customer. Nowadays a sales representative can carry a video recording in his or her brief case recorded in the language of the customer.

However while video recordings are ideal for use with relatively small groups of people, it cannot at present compete with ciné-film for large-scale presentations such as the launch of a new car where cinema quality large screen viewing is required.

Having reviewed the systems available for illustrating presentations, let's now look at the sort of material for which the various systems can be used. We will start by considering what we can show on OHP transparencies. These can be either hand drawn or computer generated.

Overhead projector transparencies

Graphs

Graphs are used to represent numerical data in pictorial form. The data is represented by a line (curve) varying in position between two axes. The horizontal axis is usually called the X-axis or the *independent* variable axis. The vertical axis is usually called the Y-axis or the *dependent* variable axis. This is because the value of Y is related to and *depends* in some way on the value of X. Therefore only two values can be represented on a simple graph. The scale of the axes can influence the appearance of the graph. They can be arranged either to exaggerate the changes being depicted or they can be arranged to minimize the apparent changes. It all depends upon how we wish to influence our audience. Let's now consider the graphs shown in Fig. 2.4.

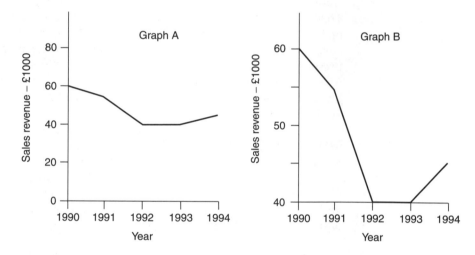

Figure 2.4 Graphs.

It is the first impression that people tend to remember. So by changing the scales we can influence that first impression. Mathematically, both Fig. 2.4(a) and Fig. 2.4(b) are identical. However, at first sight, the sales figures shown in graph (a) appear to remain reasonably constant with a slight downward trend but rising again after 1993. On the other hand, graph (b) shows what appears to be a catastrophic fall in sales. However, remember that both graphs are numerically the same. This distortion of the scales is widely used by the media to emphasize a particular point in order to influence public opinion. The use of colour can also affect interpretation. Bright colours such as red against a paler pastel coloured background can be used to focus the viewer's eye on a point of emphasis.

Multiple line graphs (compound graphs) can be used to compare related data as shown in Fig. 2.5. Here we can see how the various properties of plain carbon steels are affected by the variation in carbon content (steel is an alloy of iron and carbon). Reproduced in black and white, as in this book,

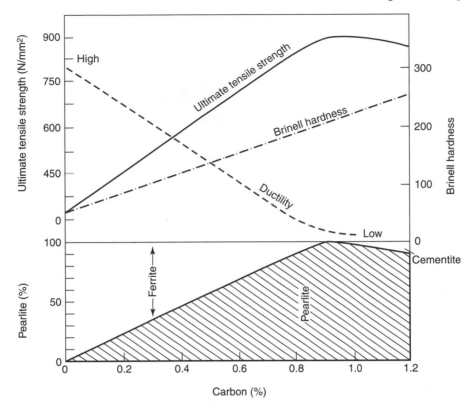

Figure 2.5 Multiple line graphs.

different types of line can be used for the different properties. Reproduced in colour, the different properties could be colour coded in contrasting colours.

Pie charts

Pie charts are one of the simplest ways of representing numerical data, as they are easy to draw and understand. The numerical value is represented by the angle of the segment – for example, if the total sales for a range of goods is £1000 the whole circle (360°) represents £1000. If sales of garment 'A' amounted to £200 over the same period of time then its segment would have an angle of 360° × (£200 ÷ £1000) = 72°. This is shown in Fig. 2.6.

The sales data originally shown using the line graph in Fig. 2.4 will now be shown as the pie chart in Fig. 2.7. To make the different segments of the pie chart stand out clearly they can be printed in contrasting colours. Again, first impressions are important and emphasis can be drawn to a particular segment by giving it a brighter and stronger colour than those segments whose importance it is desirable or advisable to play down.

Bar charts

Bar charts show numerical data by changing the heights of the bars when they are drawn vertically or the lengths of the bars when drawn horizontally. Typical bar charts are shown in Fig. 2.8. These charts show the same

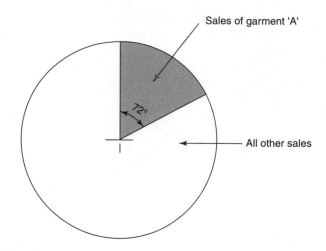

Figure 2.6 Principle of the pie chart.

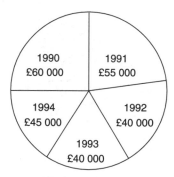

Figure 2.7 Pie chart showing sales revenue from 1990 to 1994.

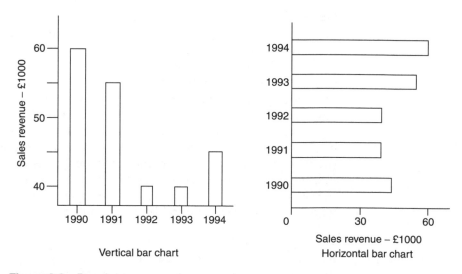

Vertical bar chart

Horizontal bar chart

Figure 2.8 Bar charts.

numerical data as Fig. 2.4 and, again, the origin of the dependent variable can be set differently to stress a point of view.

Gantt chart

The *Gantt chart* is widely used in business to show expected performance against actual performance. It is a sort of horizontal bar chart and an example is shown in Fig. 2.9(a). By having more sections, it would be possible to compare the results of two or more salespersons against each other as well as against the company sales targets, as shown in Fig. 2.9(b).

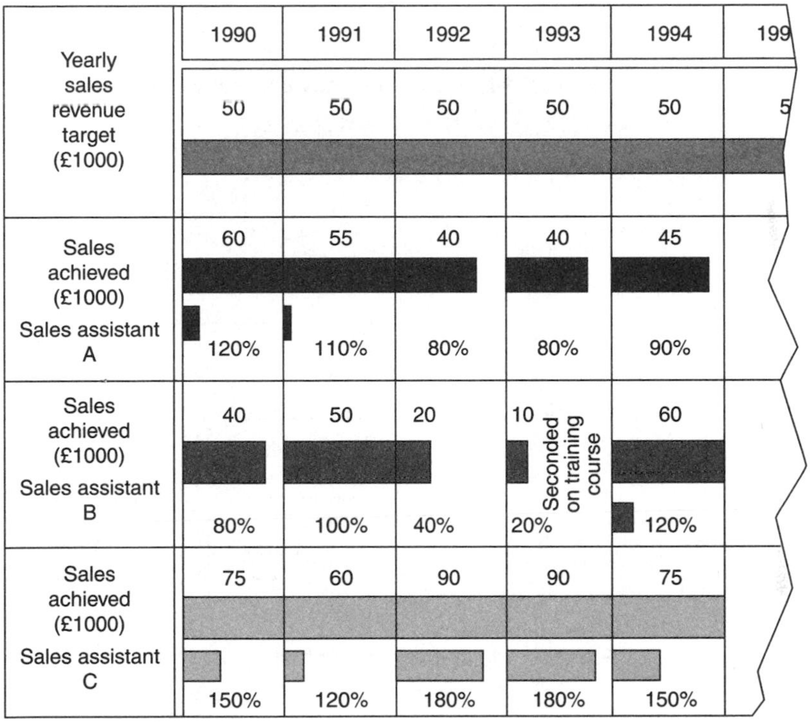

	1990	1991	1992	1993	1994	199
Yearly sales revenue target (£1000)	50	50	50	50	50	5
Sales achieved (£1000) Sales assistant A	60 120%	55 110%	40 80%	40 80%	45 90%	
Sales achieved (£1000) Sales assistant B	40 80%	50 100%	20 40%	10 Seconded on training course 20%	60 120%	
Sales achieved (£1000) Sales assistant C	75 150%	60 120%	90 180%	90 180%	75 150%	

Figure 2.9 The Gantt chart.

Pictograms (ideographs)

The pictogram is another method of representing numerical data often used when communicating with the general public and other non-technical groups. An *icon* is used to resemble the dependent variable – for example, Fig. 2.10 shows our original sales figures in the form of a pictogram. In this example the icon is the £ symbol and represents £10 000. Lesser amounts are shown by using only a fraction (part) of the icon. Pictograms are intended for giving an overall impression rather than for detailed analysis.

Charts

These are used to show the relationship between non-numerical data such as the organization chart shown in Fig. 2.11. This shows the relationship between the various departments in a company.

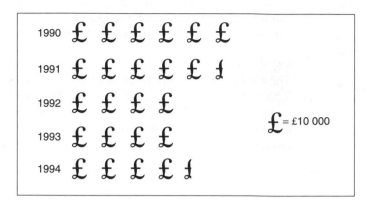

Figure 2.10 Pictogram showing yearly sales revenue.

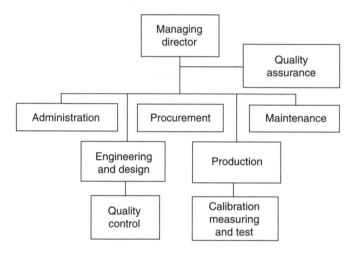

Figure 2.11 Organization chart.

Flow charts

These are logical diagrams showing the direction of flow required to determine a sequence of information. If you want to make an appointment with your doctor, you ring up your doctor. The request *flows* from you to your doctor. The request for an appointment does not start by flowing from the doctor to you. A typical example of a flow chart or diagram for a quality control system is shown in Fig. 2.12.

System flow charts

System flow charts are used where procedures can be broken down into logical steps with a simple yes/no decision. Such charts are very popular with computer systems and systems analysts since the computer works in the same logical manner. System flow charts are also known as *algorithms*. Four basic symbols are used to build up diagrams and the use of three of them is shown in Fig. 2.13.

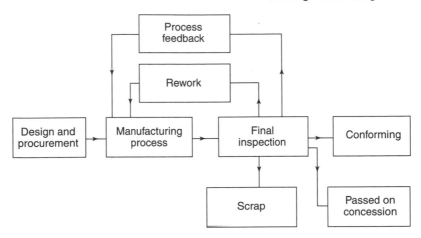

Figure 2.12 Flow chart.

Data sheets

These are used to tabulate data for easy reference. The information contained in such sheets may be distributed as single sheets, or several sheets may be bound together to make a booklet, or the data may be printed out as a wall chart. A typical example is shown in Fig. 2.14. For a given type of boiler and the number of radiators it is to feed, you can immediately see the manufacturer's recommended pipe size for coupling to the boiler. If you wish to feed 7 radiators from a type M-004 boiler, you look down the 4-8 radiator column and, opposite the type M-004 boiler, you can see that you require 15 mm diameter pipework.

When using data sheets it is important to establish that they are up to date. On receipt of a new issue, all previous data sheets should be destroyed or be returned to the point of issue. Data kept in a computer system should be on a *read only memory* (ROM) status so that unauthorized alterations cannot be made.

We have deviated slightly from our original aim of considering material for use with an overhead projector. However, selected tables from data sheets can be reproduced as OHP transparencies in order to demonstrate their use to your audience. Before we leave this section, there are a few more techniques to consider.

Photographs

Photographs in the form of colour transparencies are used to show representations of actual objects. A sequence of photographs are often taken from different viewpoints. They are useful for showing new clothing designs on live models without the expense of having the actual models present at each presentation. Large machines and structures such as bridges and oil-rigs, can be shown in their working environment.

Models

Models can be static or they can be working models. They can be scaled down or scaled up and are very useful demonstrating how a particular system or

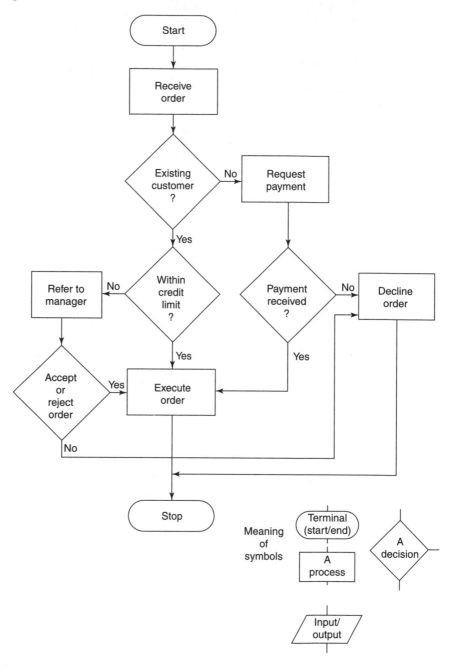

Figure 2.13 System flow chart.

mechanism works. They can be stripped down or built up in front of the audience. Make sure you are familiar with their construction and operation. Nothing looks worse than an operator fumbling around with a piece of unfamiliar equipment. It makes the audience uncomfortable and you quickly lose their confidence in everything else you have to say or do.

Pipe sizes for the range of MQ boilers				
Boiler tyre	Pipe diameter in millimetres for no. of radiators			
	1–3	4–8	9–12	13–20
M-001	4.0	5.0	6.0	8.0
M-002	5.0	6.0	8.0	10.0
M-003	6.0	8.0	12.0	14.0
M-004	10.0	15.0	20.0	25.0
M-005	15.0	20.0	30.0	40.0

Data sheet no. MQ../123
Issue no.3
Date November

Figure 2.14 Data sheet.

Handouts

Handouts should be well prepared and should be identical with your presentation. They save the audience from having to make notes under dimmed light conditions. When used as a teaching or training aid, they should be 'gapped' so that the audience have to complete them. Audience participation is an essential part of the learning process, and active participation also relieves any monotony. When handouts are for use with a visiting audience they should be smartly bound with the cover bearing the title of the contents and the company logo.

Feedback

Feedback should be by means of forms that are well constructed with unambiguous questions and easy to complete by means of boxes that can be ticked or by the use of one word answers. They should be used at the end of a presentation and should enable the person completing the form to give both quantitative and qualitative assessments.

Test your knowledge 2.6

1. State the visual aid you would use in the following circumstances and give the reason for your choice:

   (a) A prototype washing machine.
   (b) An electronic circuit diagram.
   (c) A flow chart for a manufacturing process.
   (d) A new machine in operation.
   (e) A question and answer feedback session.

2.2.2 Technical drawings

Having agreed the design with the customer it is necessary to produce working drawings of the product and the tooling that is to be used in its manufacture. Drawings may also be needed:

- to accompany sales literature;
- for installation and commissioning;

- for maintenance;
- for repairs.

The type of drawing used will depend upon which of the above applications it is intended for. Technical drawings for manufacture should be produced in accordance with the appropriate national standard – for example, mechanical engineering drawings in the UK are produced in accordance with BS 308 conventions. Examples of the drawings used by engineers for showing three dimensional solids on a two dimensions sheet of paper or on a computer screen are shown in Fig. 2.15. They are referred to as *orthographic* drawings and can be presented in:

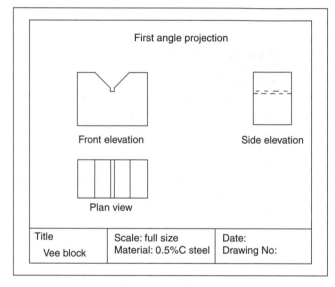

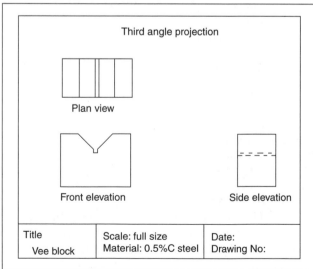

Figure 2.15 Illustration of first angle and third angle orthogonal drawings.

- First angle or English projection (Fig. 2.15(a));
- Third angle or American projection (Fig. 2.15(b)).

In addition, pictorial views are often used, particularly for DIY applications where the user may not be familiar with the interpretation of orthographic drawings shown in Fig. 2.15. The most commonly used pictorial drawing techniques used are:

- *Isometric* drawing where all the lines are drawn true length and the *receding lines* (the lines appearing to be 'going into' the page) are drawn at 30° to the horizontal as shown in Fig. 2.16(a).
- *Cabinet oblique* drawing where only the vertical lines are true length and the receding lines are half the true length. The receding lines are drawn at 45° to the horizontal as shown in Fig. 2.16(b).

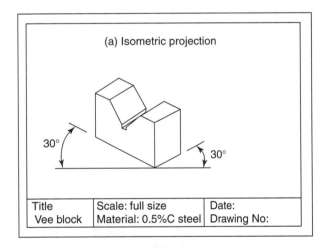

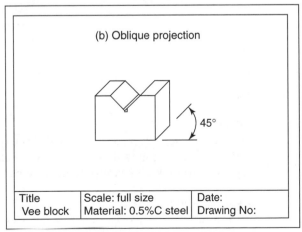

Figure 2.16 Pictorial drawing.

General arrangement (GA) drawing

Figure 2.17 shows a typical general arrangement (GA) drawing. You can see that this drawing:

- shows all the components in their assembled positions;
- lists and names all the components and lists the reference numbers of the detail drawings needed to manufacture them;
- states the quantity needed of each component;
- lists the 'bought-in' components and indicates their source;
- lists any modifications and corrections.

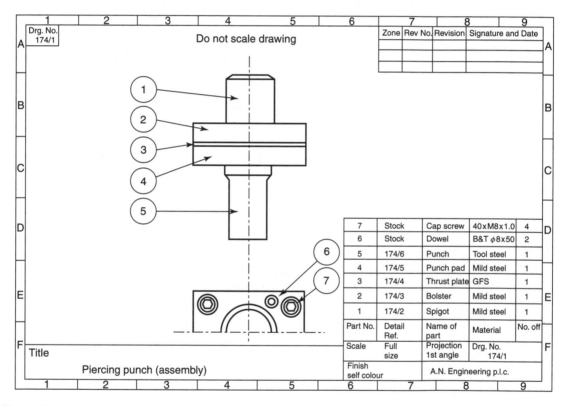

Figure 2.17 Assembly drawing.

Detail drawings

As an example, Fig. 2.18 shows one of the detail drawings listed in Fig. 2.17. A detail drawing gets its name from the fact that it provides all the *details* necessary to make the component shown.

Exploded drawing

An example is shown in Fig. 2.19. This type of drawing is widely used in maintenance manuals. It shows the parts correctly positioned relative to each other but spaced apart so that the parts are easily recognizable. Generally

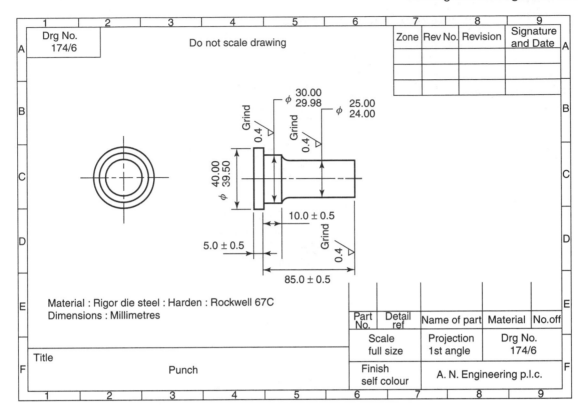

Figure 2.18 Detail drawing.

a table of parts (parts list) is included giving the necessary information for ordering spares. Similar drawings are used for the assembly of 'flat-pack' items of domestic equipment and furniture purchased from DIY stores. It makes identification of the parts easy and shows the order in which they should be assembled.

Circuit diagrams

These are drawn using standard symbols for the various components. A typical example is shown in Fig. 2.20(a). This is an amplifier circuit. Nowadays most circuits are built up on a *printed circuit board* (pcb) and, as well as the circuit, the designer also has to provide a suitable layout for printing the circuit board. Fig. 2.20(b) shows a suitable layout for the amplifier circuit.

Test your knowledge 2.7

1. Figure 2.21 shows a simple object in first angle projection. Redraw it in:

   (a) third angle projection;
   (b) isometric projection;
   (c) cabinet oblique projection.

2. Figure 2.22 shows an exploded view of a simple assembly. Redraw it as assembled ready for use.

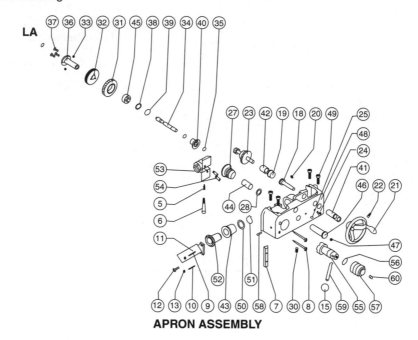

APRON ASSEMBLY

SECTION LA
APRON ASSEMBLY

Drg. Ref.	Part No.	Description	No. Off/Mc.	Drg. Ref.	Part No.	Description	No. Off/Mc.
LA5	A4729	Spring – Leadscrew Nut – – – –	1	LA38	A9782	Washer – Drive Shaft – – –	1
LA6		Cap Hd. Screw – Leadscrew Nut (2 B.A. $\times 1\frac{3}{4}$'')	1	LA39		Circlip – Drive Shaft (Anderton 1400 – $\frac{5}{8}$'')	1
LA7	A2082	Glb Strip – Leadscrew Nut – – – –	1	LA40	A9280	Knob Operating Spindle – – –	1
LA8	A9193	Ch. Hd. Screw – Strip Securing – – –	2	LA41	A9210	'Ollite' Bush – – – – –	2
LA9	A9194	Adjusting Screw – Glb Strip – – –	1	LA42	A9211	'Ollite' Bush – – – – –	1
LA10	A9195	Adjusting Screw – Glb Strip – – –	1	LA43	A9212/1	'Ollite' Bush – Flanged – – –	1
LA11	A9196	Leadscrew Guard – – – – –	1	LA44	A7595	'Ollite' Bush – – – – –	1
LA12		Hex. Hd. Set Screw (2 B.A. $\times \frac{1}{2}$'') – –	1	LA45	A9220	Clutch Insert – – – – –	1
LA13		Hex. Locknut (2 B.A.) – – – –	2	LA46	A9203/1	Stud – Gear Cluster – – – –	1
LA15	80002	Ball Knob (KB5/100) – – – –	1	LA47	65001	Oil Nipple (Tecalemit NC6057) – –	1
LA18	A9198	Hand Traverse Pinion – – – –	1	LA48	10025/1	Apron Assembly (Includes LA41, LA42, LA43)	1
LA19	65004	Sealing Plug – Apron (AQ330/15) – –	1	LA49		Cap Screw (M6 $\times 1 \times 25$ mm) – – –	4
LA20	70002	Woodruff Key (No. 404) – – – –	1	LA50	10217	Thrust Washer – – – – –	1
LA21	A2087	Handwheel Assembly – – – –	1	LA51	10431	Circlip – – – – – –	1
LA22		Socket Set Screw ($\frac{1}{4}$'' B.S.F. $\times \frac{1}{4}$'') (Knurled		LA52	A9200/1	Bevel Pinion – – – – –	1
		Cup Point) – – – – – –	1	LA53	A1975/3	Leadscrew Nut – – – – – set	1
LA23	A9199	Rack Pinion Assembly – – – –	1	LA54	10508	Cam Peg – – – – – –	2
LA24	A2531	Oil Level Plug – – – – –	1	LA55	10528	Cam – – – – – – –	1
LA25	65000	Oil Nipple (Tecalemit NC6055) – – –	1	LA56	65007	'O' Ring (BS/USA115) – – – –	1
LA27	A9201	Bevel Gear Cluster Assembly (Includes LA44)	1	LA57	10529	Eccentric Sleeve – – – – –	1
LA28	A9202	Thrust Washer – – – – –	1	LA58		Socket Set Screw ($\frac{3}{14}$'' B.S.F. $\times \frac{3}{8}$''	
LA30		Socket Set Screw ($\frac{1}{4}$'' B.S.F. $\times \frac{1}{2}$'') (Knurled				Half Dog Point) – – – –	1
		Cup Point) – – – – – –	1	LA59	10530	Lever – – – – – – –	1
LA31	A9204	Clutch Gear Assembly (Includes LA45) –	1	LA60		Socket Set Screw (2 B.A. $\frac{1}{4}$'', Cup Point)	1
LA32	A9205	Drive Gear – – – – – –	1	LA61	10424	Guard Plate (not illustrated) – –	1
LA33	73010	Ball – Clutch (S mm ϕ) – – –	2				
LA34	A9206	Operating Spindle – – – –	1				
LA35		Circlip (Anderton 1400 – $\frac{3}{8}$'') – –	3				
LA36	A9207	Drive Shaft – – – – – –	1				
LA37		C's'k Hd. Socket Screw (2 B.A. $\times \frac{1}{2}$'') – –	3				

Figure 2.19 Exploded view and part list. Courtesy of Myford Ltd.

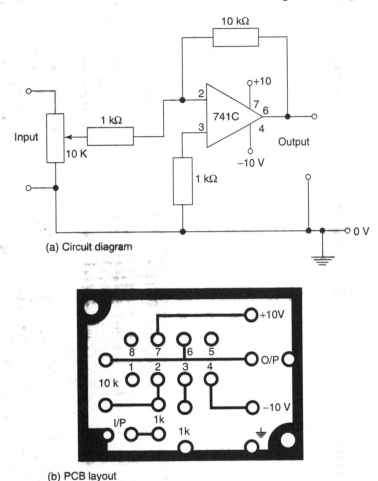

(a) Circuit diagram

(b) PCB layout

Figure 2.20 Amplifier circuit.

2.2.3 Presentation techniques

Your presentation technique must be organized to suit your target audience. A presentation suitable for your fellow professionals using the appropriate technical vocabulary would be largely meaningless if given to members of the general public. Similarly, a presentation given to the engineering design team at your firm would be unsuitable for representatives of the accounts department. Not because they are less knowledgeable but because they would require totally different information. It is also important to mix your delivery techniques to avoid monotony. This has been mentioned earlier, when introducing the audio-visual techniques available. Let's now consider the techniques available to the presenter.

Verbal

Rehearse your presentation thoroughly so that you do not have to read it from the script. When discussing speaking notes earlier in this chapter it was

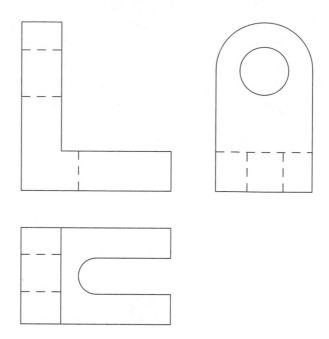

Figure 2.21

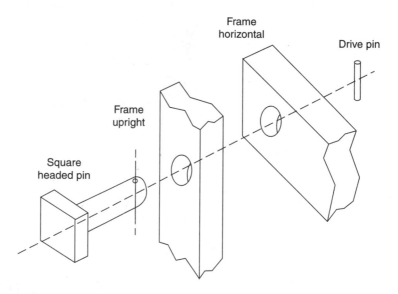

Figure 2.22

recommended that you try to use the headings as prompts. The details are only there in case you have a lapse of memory.

- Don't use words you find difficult to pronounce.
- Try to be enthusiastic and let your voice show it.
- Be in command of your subject. The less you have to refer to your notes the more impressed will be your audience.

- Speak up so that everyone can hear, and speak clearly. Don't rush, pause at the end of each sentence for a moment.
- Don't hesitate and don't repeat yourself.
- Don't fidget with your hands and don't pace about. Have all your audio-visual aids organized conveniently so as to keep movement to a minimum.
- Arrange the verbal part of your presentation so that there are 'natural breaks' where you can introduce your audio-visual material without interrupting your flow and spoiling the continuity. Remember that most people's concentration span is quite short.

Overhead projector

The overhead projector and its transparencies have already been discussed. Now let's see how to use it to the best advantage.

- Make sure to set it up properly before the start of your presentation so that it is the correct distance from the screen. The whole of your transparency should be visible without material near the edges falling off the screen. Mounting your transparencies in cardboard frames helps to avoid this happening since the projected area is a constant size.
- Make sure the projector is focused before you start. A coin placed on the illuminated panel provides a sharp outline for focusing.
- Don't leave the projector on all the time. Only turn it on when you need it. To leave it on distracts the attention of your audience from you, the speaker. Remember it's your presentation. You're the star of the show.
- If there is more than one piece of information on the transparency don't show it all at the same time. This can be distracting and confusing. Cover the transparency with a sheet of thin card and reveal each item of information as and when you require it.
- Remember that if the audience want to reinforce a particular piece of information you can easily return to an earlier transparency. For this reason make sure all the transparencies are numbered sequentially, the numbers relate to the prompts in your notes and that you stack transparencies in the correct order as you remove them from the projector.
- Overhead projectors can be used with other devices apart from transparencies. You can obtain measuring instruments such as voltmeters and ammeters with transparent scales, and transparent, liquid crystal computer screens that can be placed on an OHP and used in conjunction with CAD software, databases and spreadsheets.
- Finally, remember that you can obtain hard copy from your transparencies and computer displays. Such handouts remove the need for your audience having to make notes in dimmed light conditions and enable them to concentrate on your presentation. If you have handouts available, make this clear to your audience at the start. Not only will it be too late after they have struggled to take notes, you do not want them leaving your presentation feeling annoyed at this simple oversight.

Slides

The use of 35 mm transparencies were discussed earlier in this chapter. They are useful for presenting real-life images of 3D objects in colour. Make sure that they are strictly relevant and not just 'pie-filling'. Also make sure they are loaded into the magazine in the correct sequence and the correct way up!

Chalkboard

This should not form part of your main presentation but is useful for quick diagrams and notes during a question and answer session following the main presentation. Experience and practice are required to draw and write neatly on a chalkboard.

Flip chart

This is a large pad of white paper supported on an easel. Felt pens of various colours and line thickness are used to draw or write on the paper. It is normally used under the same circumstances as a chalkboard and the same comments apply. It has the advantage of being cleaner (no dust) and, when full, you merely turn the paper over and continue on the next sheet. You can, therefore, turn back to an earlier sheet whenever the need arises. With a chalkboard, it has to be cleaned off when full and the material on it is lost for ever.

Video recording

An excellent method of presenting material. Unfortunately, although it is easy to make a video recording with modern lightweight camcorders, it is very difficult to make good quality and properly edited recordings. We are so used to watching high quality material on television that, unless the recordings used in your presentation approach these professional standards, they are best not used at all. This contrast between amateur and professional recording is shown up clearly in television programmes that make use of home-video material sent in by viewers.

Written

Written material has already been discussed in this chapter. It should be used for two purposes:

- as a prompt for the presenter;
- as back-up handouts to the audience so that they do not have to take notes in dimmed light conditions.

Test your knowledge 2.8

1. Explain briefly what steps you would take to make a verbal presentation as professional as possible.

2. Explain how you could illustrate your presentation with easily produced visual aids.

3. Explain what steps you would take to use an OHP effectively.

4. Tabulate the advantages and disadvantages of flip charts compared with a chalkboard.

2.2.4 Glossary (technical vocabulary)

The technical vocabulary used should be appropriate for your audience and for the product or design you are promoting. Your audience will often contain persons of:

- different technical and professional backgrounds;
- different social and educational backgrounds;
- different ethnic groups for whom English may not be their first language.

If you are presenting a new technology, the jargon terms associated with it may not yet have filtered through into general usage even among professional persons. Therefore it is often useful to issue a glossary of the terms used in your presentation. This applies particularly to *acronyms*. Acronyms are words made up from the initial letters of other words, for example NATO which stands for the North Atlantic Treaty Organization. Table 2.3 lists a technical vocabulary and some common acronyms that may be helpful in making a presentation.

Table 2.4 Glossary

Technical word or acronym	Definition
A	
ABS	Antilock braking system.
ABS	Acrylonitrile-butadiene-styrene: a tough, high impact resistant plastic material.
a.c.	Alternating current (electricity).
Adhesive	A substance for joining materials together by bonding.
Algorithm	A set of precise rules which specify a sequence of actions. Often presented in the form of a flow chart.
Alloy	A complex metal consisting of a base metal plus other metal and/or non-metals in close association.
Alternator	An electromechanical device for producing alternating current (electricity).
Ampere (amp)	Unit of magnitude of electrical current.
Analysis	A detailed examination to determine the constituent parts.
Appearance	How a thing looks.
Assembly	A product made up of many individual parts. The act of putting parts (components) together.
B	
Bakery	An establishment where food products are manufactured by a baking process, e.g. bread, cakes, biscuits.
Batch production	A production system where medium and large numbers of identical products are manufactured at a single setting of the machines.
Battery	A number of electrical cells connected together in series to increase the terminal voltage, or in parallel to increase the current handling capacity.
Binary	A base 2 numbering system used in computers and other digital devices.
Board	A committee of the directors representing the shareholders of a company and responsible for determining its overall policy.
Brewery	An establishment for the manufacture of ales, lagers and stouts by the process of brewing.
Budget	Money allocated for a particular project, thus setting a restriction on the amount that can be spent on that project.
C	
CAD	Computer aided design.
CAM	Computer aided manufacture.
Characteristic	A distinguishing property or quality.
Complexity	The intricacy of a problem, system or mechanism.

(*Continued overleaf*)

Table 2.4 (*Continued*)

Technical word or acronym	Definition
Components	The individual constituent parts of an assembly.
Constraints	The restrictive conditions imposed on an activity.
Construct	Assemble or build.
Conventional	Customary, usual or traditional.
Criteria	The standards or specifications by which something can be judged.
D	
Database	A system for storing information, usually on a computer.
Demand	A need or requirement for a product.
d.c.	Direct current (electricity).
Development	The act of development or growth as applied to the improvement of a design or the size of a company.
Director	A person appointed to the board of a company to represent the shareholders' interests and assist in the proper running of the company.
DOS	Disk operating system as applied to a computer.
Dynamo	An electromechanical device for the production of direct current (electricity).
E	
Electrical engineering	A branch of engineering concerned with the generation of electricity, and the design and manufacture of apparatus powered by electricity.
Electronic engineering	A branch of electrical engineering involving small current devices controlled by solid state devices such as diodes, transistors and integrated circuits.
Ergonomics	The study of the relationship between human beings and their physical (working) environment.
Evaluation	The assessment of data or the process of calculation.
F	
Feedback	The response of a target audience to a presentation. Any system where part of the output is fed back to the input.
Finance	The management of money. Money resources.
Flow production	The manufacture of identical or similar products on a continuous basis so that the work flows through the factory in an unbroken stream.
Function	The purpose of an object or system. The act of operating or performing as specified.
G	
GDP	Gross Domestic Product.
GNP	Gross National Product.
Generator	An electromechanical device for producing electricity.
H	
Histogram	A type of graph where numerical data is represented by vertical bars.
Home market	A market for the sale of goods in the country of their origin.
I	
Initiate	To take the first step, to set going:
Integration	Combining together or blending.
Integrity	Honesty, soundness.
J	
Jobbing (workshop)	A workshop that makes single items and small batches of components or assemblies to order.
Joinery	The manufacture of wooden devices such as doors, window frames and staircases. A factory where such things are manufactured.
L	
Loss	A loss is made when the selling price of a product is less than the price at which it can be sold.
M	
Machine	A device or system for transferring energy from one form to another in order to do mechanical work.

Table 2.4 (*Continued*)

Technical word or acronym	Definition
Maintenance	The process of keeping structures and equipment in proper order.
Marginal	Close to a prescribed limit.
Marketing	The process of identifying and analysing the requirements of a market and providing the publicity necessary to support the sales organization.
Modification	An adjustment or alteration.
O	
Option	Choice.
Output	The amount produced in a given time.
P	
Performance	The functional quality of a device or system.
Pollution	The contamination of the environment by noise, unsightly constructions and harmful substances.
Process	A sequence of operations.
Prototype	An initial model produced for trials and testing purposes.
Q	
Quality	Fitness for purpose, standard of excellence.
R	
RAM	The random access memory of a computer. Data can be written to this memory or retrieved as required. It can also be deleted when no longer required.
Reliability	The ability to function over long periods of time without failure.
Research	To investigate.
ROM	The read only memory of a computer. The data stored in this memory can be accessed but not altered.
S	
Specification	A detailed description, e.g. of work to be done, or the requirements of a design, or the quality of workmanship required.
Standard	An accepted approved example against which others are judged or measured.
Synthesis	Putting together, building up from elements.
System	A set of related components. A body of organized concepts.
T	
Tolerance	The amount by which it is permissible for a dimension to vary.
V	
Volt	Unit of electrical potential.
W	
Watt	Unit of electrical and mechanical power (rate of doing work).
Weave	The process of manufacturing cloth on a loom.
Weight	The effect of the earth's gravitational force acting on the mass of a body.

2.2.5 Feedback

In order to finalize the design, it is necessary to obtain customer feedback at the end of the presentation. Two levels of assessment (judgment) have to be made.

Quantitative judgments

These relate to *measurable features* of the design such as size, weight, performance, strength and cost. In the case of foodstuffs judgments may be made relating to such dietary matters as fat and sugar content.

Qualitative judgments

These relate to such matters such as colour, finish and fitness for purpose, that is, the quality of the finished product and whether it represents value for money for the customer.

Once the client has carried out their analysis of the proposed design, user panels are frequently set up across the country to assess customer reaction. User panels consist of small groups of people of selective age and sex groups. They are given the opportunity to try out prototype samples of the design and comment upon it. They may also be invited to comment upon the proposed packaging and advertising literature. This information provides feedback for the manufacturer, the marketing personnel and the design team. Final amendments are then made to the design, packaging and advertising literature and the design is put into production. Figure 2.23 shows a simple feedback report form.

Product	Design Team				
Functions	Performance				
	U	M	A	G	O
Operating instructions					
Ease of operation					
Packaging					
Advertising					
Material					
U = Unacceptable M = Marginal A = Acceptable G = Good O = Outstanding					
General comments					

Figure 2.23   Feedback form.

Test your knowledge 2.9

1. Explain the importance of feedback as part of a presentation.

2. Explain the essential differences between quantitative judgments and qualitative judgments.

3. Draw up a suitable feedback form to be used in your Evidence indicator activity 2.2.

2.2.6 From design brief to finished product

Finally, let's now consider a simple product from the initial request to the design team to the finished product.

Design brief

The client is a furniture manufacturer producing popular low cost furniture items for the mass market. Market research has shown that there is a demand for a simple, low-cost stool to seat one person.

- It is to be made from wood.
- It is to have a conventional polished finish.

- It is to be light in weight but stable.
- Any cushion is to be supplied as a separate item from a specialist supplier.
- It must be strong enough for a heavy man to stand on while changing a light bulb or reaching to get articles down from a high shelf.
- It must be simple to manufacture and competitively priced.

Design proposal

Two designs were prepared. One was for a traditional bench type stool as shown in Fig. 2.24(a) and one for a more modern concept in laminated wood as shown in Fig. 2.24(b). It was felt that the second design might be acceptable in view of the fact that the manufacturer had the technical resources and experience to work in this mode. Comparative, specimen costings were produced for both designs. The second design saved on assembly costs but required higher tooling costs to set it up. In view of the fact that large batch production would be employed if the market research was correct, there should be no difficulty in recovering the set-up and development costs.

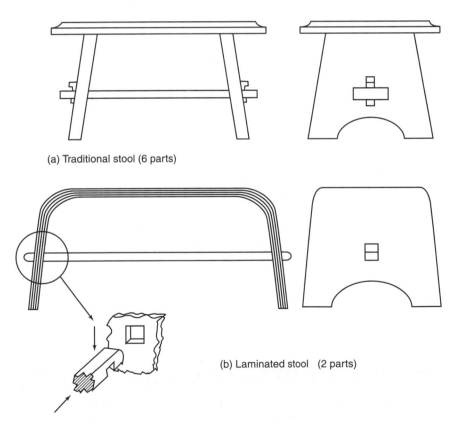

(a) Traditional stool (6 parts)

(b) Laminated stool   (2 parts)

Figure 2.24   Design proposals for a stool.

Presentation

This was kept quite simple as only a small audience would be present consisting of representatives of the company commissioning the design proposal. Present were the managing director, the chief accountant, the marketing manager and the production manager.

A verbal presentation was given by the leader of the design team. Prototypes of the stools were available as visual aids for examination and a cost analysis was presented with the aid of an OHP. The presentation was supported by a printed handout of all the presentation material used together with copies of the manufacturing drawings.

Feedback

The client company congratulated the design team on their work. After discussion it was decided to adopt the second design (Fig. 2.24(b)) for the following reasons.

- The design was in the modern idiom and unlike anything being offered by the competition.
- The design was easy to produce with few parts and little assembly.
- The rounded form and smooth lines would make it easy to maintain and keep clean.
- It satisfied all the key features of the design brief.

Subsequently, however, the client company pointed out that they had a *legal duty of care* to their customers and having carried out a *risk assessment* it was considered that the rounded ends could cause the user to slip and fall. The risk would be greater if the user was standing on the stool. The fact that the stool was not intended to be stood upon would not exempt the manufacturer from being sued in the event of an accident. Also it was found that the shape made it difficult to keep a cushion in place.

Finished product

The design team accepted the comments as fed back to them and revised the design as shown in Fig. 2.25. By adding a top 'tray' to the design, the risk of slipping off the rounded ends was removed. The 'tray' also helped to keep the cushion in place. Cutouts were incorporated in the ends of the 'tray' as shown and these made useful carrying handles. As with the prototype design, modern high strength adhesives were used throughout and there were no metal fastenings required. It was considered that the extra cost involved in the extra work required for the final design was worth while in view of the fact that it resulted in a safer and better product.

Key Notes 2.2

- Reports are a written or spoken (verbal) method of making a presentation.
- Reports need to be drawn up in a logical manner so that they are easily understood.
- The language used should be suitable for the target audience.
- Notes are personal to the speaker and should only be used as an aide mémoire. Avoid reading from your notes. They are only there to prompt you if your memory momentarily fails.
- Overhead projector (OHP). This is an optical device for projecting transparencies drawn or printed on acetate sheets onto a screen.

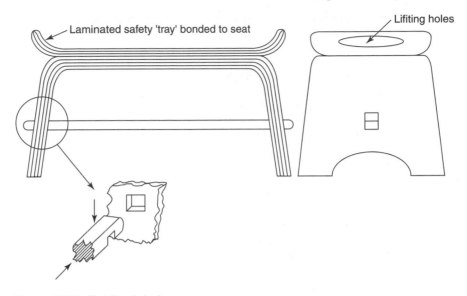

Figure 2.25 Finalized design.

- 35 mm transparencies are colour photographs taken with an ordinary 35 mm camera and which are suitable for use with a slide projector.
- Video recordings are taken with a 'camcorder' and are suitable for showing on a television set. They can produce moving or still pictures in colour and sound is also available.
- Graphs are a means of showing numerical relationships and data in pictorial form.
- Pie charts are also a simple type of graph much used in presenting numerical information to non-technical audiences. They are frequently seen in the newspapers.
- Bar charts (histograms) use vertical bars or columns to show numerical data.
- Gantt charts are also a form of bar chart but the bars are arranged horizontally. They are often used to compare actual performance with expected performance.
- Pictograms (ideographs) are a pictorial method of presenting numerical data for non-technical persons. Icons are used to resemble the dependent variable.
- Charts are used to show non-numerical data such as company organization and relationships.
- Flow charts are used to show the direction of flow required to determine a sequence of information.
- System flow charts are used where procedures can be broken down into logical steps with a simple yes/no decision. They are widely used by computer systems analysts. System flow charts are also called *algorithms*.
- Data sheets are use for presenting numerical data for easy reference – for example, inch to millimetre conversions.
- Models can be static or working. They can be scaled-up versions of very small mechanisms that are difficult to see, or scaled down versions of very large mechanisms that are too large to bring into the room being used for a presentation.
- Handouts are a printed record of the presentation for the audience to take away with them. They save having to take notes under dimmed light conditions.
- Feedback should be by well-constructed forms issued at the end of a presentation. They are essential for design assessment.
- Technical drawings are used to communicate with the manufacturer of the product at the production management and shop floor level. They are also used by customers for installation and maintenance purposes.
- Technical drawings produced in the UK are drawn to BS 308 conventions.

- Drawings may be orthographic and in first angle or third angle projection, or they may be pictorial and in isometric or oblique projection. Exploded views are used for DIY assembly drawings and machine maintenance drawings.
- General arrangement (GA) drawings show all the parts of an assembly in their correct positions and list the detail drawings needed to manufacture the components used in the assembly together with any 'bought-in' parts required.
- Detail drawings show only one component on each drawing together with all the information needed to manufacture that component.
- Presentation techniques vary depending upon the target audience. A combination of spoken and audio-visual techniques should be used to capture and maintain the attention and interest of the audience. These require much pre-planning and practice.
- Chalkboards and flip charts should only be used for discussion sessions, not for the main presentation.
- Glossary (technical vocabulary). This is used to explain the meaning of unfamiliar technical (jargon) terms and acronyms used during the presentation. Acronyms are words made up from the initials of other words, for example UNO (United Nations Organization).
- Feedback is required in order to judge the reception the audience has given to the presentation. In the case of a design proposal, the feedback is necessary to finalize the design ready for manufacture.

Evidence indicator activity 2.2

Prepare a presentation for both the design proposals you used in Evidence Indicator activity 2.1. The support material should be an appropriate mixture of verbal, written and visual material. Do not be overambitious. Simple material well presented is more satisfactory than a complex presentation that is badly prepared.

At the end of your presentation use the appropriate quantitative and qualitative feedback techniques.

Finally, in the light of the audience feedback obtained, show how the final design was modified ready for manufacture.

<table>
<tr><td>**Chapter**
3</td><td># Production planning, costing and quality control</td></tr>
</table>

Summary	This chapter covers the requirements of Unit 3 of the GNVQ Manufacturing (intermediate). The *unit* is subdivided into three *elements*. The first element (3.1) is concerned with the production planning. The second element (3.2) is concerned with the calculation of the cost of a product. The third element (3.3) is concerned with quality assurance.

3.1 Production planning

3.1.1 Production planning

Production planning is an essential organizational exercise that is carried out before commencing manufacture. We do this regularly as part of the organization of our lives. You may decide that you want a cooked breakfast in the morning. It is no good waiting until the morning comes and find that you do not have suitable ingredients or equipment. So, you think ahead.

- Transfer bacon and sausages out of the freezer and into the fridge so that they can thaw out for the morning.
- Check that the frying pan, spatula and plate are clean and ready for use.
- Check that you have eggs, tomatoes, mushrooms, suitable cooking oil and bread.
- Is the tea, milk, sugar, teapot, and cup and saucer ready? You can now go to bed in the sure knowledge that, barring a power cut, you are all set for your cooked breakfast.
- Estimate the cooking times and set your alarm clock so that you have time to cook and eat the meal before setting out for work, that is, *schedule* your start to the day.

You have planned the production of your breakfast. You have thought ahead. This is production planning on a small scale but, in principle, it is exactly what has to happen in industry no matter whether you intend to manufacture a nut and bolt or a motor car.

Let's now see what is involved in the production of a typical industrial *production plan*. The plan will consist of a number of elements. These are listed below.

- Material sourcing and procurement so that suitable material will be available in sufficient quantity for production to commence and continue without interruption.
- The product specification.
- Analysis of the manufacturing process so that the individual tasks can be listed.
- Assign the appropriate resources to each task.
- Estimate the time that each task will take.
- Arrange the tasks in a logical sequence so that the work will flow smoothly through the factory.

- Assign starting and finishing dates and times to the individual tasks, that is, *schedule* the production.

Finally, all the above elements have to be brought together into a *production plan* which then has to be communicated to all the appropriate personnel. The production plan must be drawn up in the conventional format for the manufacturing sector concerned.

3.1.2 Product specification

A product specification can vary according the information needed – for example, a sales person would be concerned with details of size, colour and performance. A production planner is more concerned with those features that are going to influence the materials used and the manufacturing processes required. A product specification normally contains the sort of information shown in Fig. 3.1.

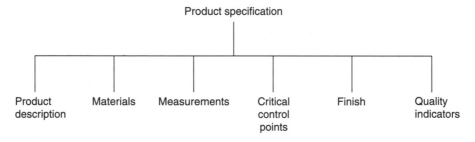

Figure 3.1 Product specification.

Figure 3.2 shows a wooden carrying box and Fig. 3.3 shows a product specification for the box.

Critical control points

These are the points when, during the course of manufacture of a product, the various components and subassemblies must be checked to make sure that they are correct to specification. This is to ensure that any defective items are rejected before they receive expensive processing or become built into the final assembly.

The critical control points for our box are as follows:

- The materials must be checked for type, size and quality before they are issued to the workshop.
- The adhesive must be checked to ensure that is the correct type for the process involved and will produce a sound joint. Sample joints should be tested for strength.
- The wood product components must be checked for quantity, size and shape after cutting.
- After gluing, power stapling and assembly the boxes must be inspected to ensure sound joints and correct assembly.

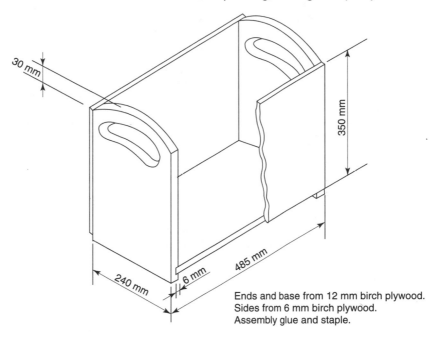

Ends and base from 12 mm birch plywood.
Sides from 6 mm birch plywood.
Assembly glue and staple.

Figure 3.2 Carrying box.

Product specification	Job number: 0078/98
Product description	Carrying box for batteries
Materials	Ends and base from 12 mm birch plywood
	Sides from 6 mm birch plywood
	Adhesive: Stickall
	Lacquer: coverite in red, yellow, green, blue, gloss
Measurements	All dimensions as drawing ±1 mm
Critial control points	1. Check materials
	2. Check sizes of blanks after cutting
	3. Check sizes after assembly
	4. Check finish after painting
Finish	Gloss lacquer in bright colours. Spray or brush on depending on quantities
Quality indicators	Materials: thickness and appearance
	Dimensions: measured for correct size
	Finish: appearance

Figure 3.3 Product specification for carrying box.

- After finishing (sanding smooth and coating with a spray lacquer), the boxes must be finally inspected for damage or defects before they are finally packed in such a manner as to protect them from damage during storage or transit.

Quality indicators

These have been introduced previously in Section 1.2.5. Quality indicators can be *variables* or *attributes*. These can be applied to the critical control points for our box as shown in Table 3.1.

Table 3.1 Quality indicators

Critical control point	Quality Indicators	
	Variables	Attributes
Materials	Check plywood sheets for correct thickness	Check plywood sheets for: (i) correct type of wood (birch) (ii) freedom from surface blemishes
Cut blanks	Check for correct dimensions as stated on the drawing ±1.0 mm	–
Assembly	Check assembled boxes for correct overall dimensions ±1.0 mm	Check assembled boxes for: (i) Correct gluing and tacking (ii) Squareness of corners (iii) Edges and joints sanded smooth
Finish	–	Check finished boxes for: (i) Freedom from blemishes (ii) Correct paint colour (iii) Correct logo/decal if required

3.1.3 Key production stages

The key stages of production that are common to all manufactured goods were introduced in Section 1.2.1. As a reminder, these are summarized in Table 3.2.

Table 3.2 Key stages of production

- Material preparation
- Processing
- Assembly
- Finishing
- Packaging

Let's now expand this table to include the key stages of production for our box. These are shown as a flow chart in Fig. 3.4. Remember there is no single correct way to make these boxes. The processes chosen and their sequence will depend upon the quantity being manufactured and the manufacturing facilities available. The processes will also reflect the personal preferences of the production manager. Figure 3.4 only shows one typical sequence as an example.

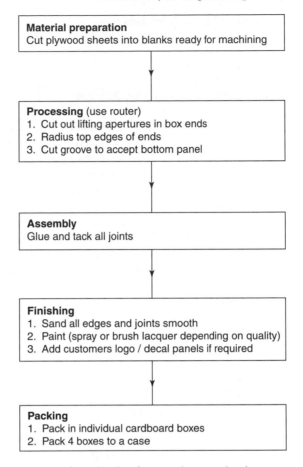

Figure 3.4 Key stages of production for wooden carrying box.

Test your knowledge 3.1

1. Explain why 'production planning' is necessary in any manufacturing industry.

2. For an article of your choice, draw up a typical 'product specification' for:

   (a) a production planning engineer;
   (b) a sales manager.

3. For an article of your choice produce a flow chart showing the key production stages.

3.1.4 Resource requirements

We have already looked at many of the resources used in manufacturing in Chapter 1. First, however, let's remind ourselves of the main resource requirements and then see how some of them can be applied to the manufacture of our box.

Capital resources

These are the resources purchased by the company for the manufacture of the company's products. They represent a major part of the company's assets.

Ignoring the premises (buildings) in which the company is housed, typical capital resources are:

- process plant;
- machinery;
- manufacturing equipment.

None of these items is *consumable* – for example, a lathe is an item of capital plant. It is a *capital resource*. The cutting tools used in it are *consumable*. They wear out and have to be resharpened or replaced frequently in the course of manufacturing the company's products. Similarly a sewing machine is also a *capital resource*. The needles wear out or break and have to be replaced from time to time. They are *consumable*.

Even capital resources wear out eventually or become obsolete but this takes place over a very long period of time compared with consumable items. Money must be put aside on a regular basis from the profits of a company for the eventual replacement of its capital resources (plant).

Tooling resources

These can be consumable items of a general nature applicable to a variety of jobs. Examples of such items are:

- cutting tools (drills, lathe tools, milling cutters, router blades, etc.);
- sewing machine needles;
- spot-welding electrodes;
- sanding belts and discs;
- polishing mops.

Tooling resources can also be *job specific* such as:

- casting patterns;
- plastic moulds;
- templates;
- machining and welding jigs and fixtures.

Although frequently very expensive, these are used only for one specific product. They are charged directly to that product and are discarded as valueless when that product is no longer made.

Material resources

These are the materials required for the manufacture of various products and will vary from one industry to another. Material resources can be broken down into:

- raw materials such as metal in bar form, forgings or castings in the engineering industry, or rolls of cloth for the clothing industry;
- standardized, finished components such as nuts and bolts in the engineering industry, or buttons and zip fasteners in the clothing industry;
- subassemblies such as electric motors, pumps and valves for the engineering industry.

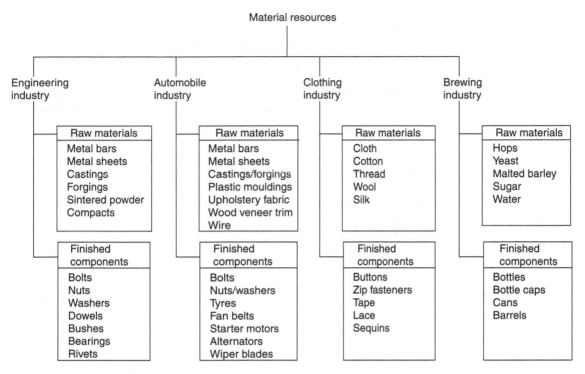

Figure 3.5 Typical material resources.

Some examples of the material resources required for manufacturing are shown in Fig. 3.5.

Human resources

Human resources refer to all the people working in the various manufacturing industries.

Direct labour refers to persons employed directly in the manufacture of a product and whose wages can be charged directly to that product.

Indirect labour refers to persons involved in the management of the company, in marketing and sales and in various clerical activities essential to the running of the company. Their wages cannot usually be charged to a specific product but have to be shared among all the products manufactured by the company.

Outsourced human resources refers to persons servicing a company but not directly employed by it. Increasingly companies are outsourcing the fringe services they require on a franchise basis rather than using employees on their own payroll – for example, the use of outside caterers to run the works canteen services. This enables costs to be more closely controlled.

Further, with the communication systems now available and the power of personal computers, it is possible for more and more persons to work from home on a freelance basis – for example, the book illustrators and copy editors who work from home for the publishing industry. Technical support and maintenance is also an area that is nowadays franchised out to factory trained individuals who are self-employed.

Service resources

These are the service resources used in manufacturing. In this context they do not include telephones, drinking water, sewerage, etc. Examples of service resources as applied to manufacturing are:

- fuel gas for heating process plant;
- process water;
- process drainage and the safe storage and disposal of hazardous liquids;
- electricity;
- compressed air.

Typical resource requirements for our wooden carrying box are summarized in Table 3.3.

Table 3.3 Resource requirements – wooden carrying box

Production stage	Operation	Type of resource	Resource required
Material Preparation	Cutting plywood sheet to size	• Material • Capital • Human • Service	• Sheets of birch plywood • Circular saw • Operator • Electricity
Processing	Forming box ends	• Capital • Tooling • Human • Service	• Router machine • Templates • Operator • Electricity
Assembly	Glueing and tacking joints	• Material • Capital • Capital • Human • Service	• Staples and glue • Powered staple gun • Hot glue applicator • Operator • Electricity
Finishing	Sanding and painting	• Material • Capital • Capital • Capital • Capital • Human • Human • Service	• Lacquer • Orbital sander • Air compressor • Spray paint gun • Spray booth • Operator for sanding • Painter • Electricity
Packaging	Packing the finished boxes	• Material • Material • Human	• Cardboard boxes • Self-adhesive tape • Packer

3.1.5 Processing times

From the key stages of production we next move on to the resources we would require to make our carrying box. However, before we can start considering the cost of manufacture, we need to work out the time taken to carry out each production stage.

$$\text{Production stage time} = \text{set-up time} + (\text{operation time per item} \times \text{number of items})$$

Using this formula we will now work out some examples. Note that we must always work in common units. Usually these are either minutes or seconds.

Example 3.1

A joiner takes 2 minutes to set up a circular saw and 15 seconds to cut each piece of wood. Calculate (a) the time to cut 10 pieces of wood, (b) the time to cut 500 pieces of wood.

(a)

$$\text{Production stage time} = \text{set-up time} + (\text{operation time per item} \times \text{ number of items})$$

$$= 2 \text{ minutes} \times 60 + (15 \text{ seconds} \times 10 \text{ items})$$

$$= 120 \text{ seconds} + 150 \text{ seconds}$$

$$= \textbf{270 seconds} \text{ or}$$

$$= \textbf{4.5 minutes}$$

We could have worked in minutes, in which case the example is as follows.

$$\text{Production stage time} = \text{set-up time} + (\text{operation time per item} \times \text{number of items})$$

$$= 2 \text{ minutes} + \{(15 \div 60 \text{ minutes}) \times 10 \text{ items}\}$$

$$= 2 \text{ minutes} + 2.5 \text{ minutes}$$

$$= \textbf{4.5 minutes}$$

The answer is the same in both cases as, of course, it should be.
Now let's work out the time for the larger quantity.

(b)
This time we will again work in minutes.

$$\text{Production stage time} = \text{set-up time} + (\text{operation time per item} \times \text{number of items})$$

$$= 2 \text{ minutes} + \{(15 \div 60 \text{ minutes}) \times 500 \text{ items}\}$$

$$= 2 \text{ minutes} + 125 \text{ minutes}$$

$$= \textbf{127 minutes} \text{ or}$$

$$= \textbf{2 hours 7 minutes}$$

Similarly, if this was part of the mass production of window frames then possibly 5000 identical pieces of wood might be required. In this case the total time taken would not be ten times greater because the setting time is the same no matter how many pieces of wood are cut. This is the time saving of large batch production over jobbing production. Try working it out for yourself, you should arrive at a time of 20 hours 52 minutes.

This might be too long a time and would slow down the overall production of window frames. If two circular saws and two operators are available then *concurrent* production can take place, that is, two saws cutting identical pieces of wood at the same time.

This time we have an additional formula.

$$\text{Number of items per operator} = \text{total number of items required} \div \text{number of operators}$$

So, if we cut 5000 pieces of wood using 2 operators then:

$$\text{Number of items per operator} = 5000 \div 2 = \mathbf{2500}$$

Reverting to our original formula the time taken would be:

$$\text{Production stage time} = \text{set-up time} + (\text{operation time per item} \times \text{number of items})$$

$$= 2 \text{ minutes} + \{(15 \div 60 \text{ minutes}) \times 2500 \text{ items}\}$$

$$= 2 \text{ minutes} + 625 \text{ minutes}$$

$$= \mathbf{627 \text{ minutes}} \text{ or}$$

$$= \mathbf{10 \text{ hours } 27 \text{ minutes}}$$

You might have thought that having two operators would have halved the overall time but you have to allow for setting up the extra machine, hence the extra 2 minutes. Table 3.4 shows how we can build up the time for manufacturing 500 of our carrying boxes.

Test your knowledge 3.2

1. Analyse the resources for your school or college under the headings considered in Section 3.1.4.

2. A machinist takes 10 minutes to set up a milling machine and then takes 5 minutes to machine each component. Calculate the time taken to produce: (a) 50 components, (b) 500 components, (c) 1000 components if two machines and two machinists are used in this particular instance.

3.1.6 Production schedules

Think of all the parts that go to make up a car. They all have to arrive at the assembly line in the correct quantities at the correct time. Think of the chaos if only body shells arrived but no engines or if only three wheels arrived for each car.

Consider the clothing trade. A factory making shirts must have regular supplies of material and buttons. If the wrong quantities arrived at the wrong time the flow of production would break down resulting in huge stocks of

Table 3.4 Time to manufacture – 500 wooden carrying boxes

Production stage	Machine set-up time (min) (A)	Number of operators (B)	Number of parts (C)	Time per operation (min) (D)	Total operation time (C) × (D)	Production stage time (A) + (C) × (D)	Aggregate time (min)
Cutting base blanks	2.0	1	500	0.25	500 × 0.25 = 125	2.0 + 125 = 127	127
Cutting end blanks	2.0	1	1000	0.25	1000 × 0.25 = 250	2.0 + 250 = 252	127 + 252 = 379
Cutting side blanks	2.0	1	1000	0.20	1000 × 0.20 = 200	2.0 + 200 = 202	379 + 202 = 581
Radiusing ends	5.0	1	1000	0.15	1000 × 0.15 = 150	5.0 + 150 = 155	581 + 155 = 736
Cutting hand grips	5.0	1	1000	0.40	1000 × 0.40 = 400	5.0 + 400 = 405	736 + 405 = 1141
Cutting base slots	5.0	1	1000	0.20	1000 × 0.20 = 200	5.0 + 200 = 205	1141 + 205 = 1346
Glueing and tacking	–	2	500 sets	3.00	500 × 3.0 = 1500	1500	1346 + 1500 = 2846
Sanding	–	2	500 assemblies	2.00	500 × 2.0 = 1000	1000	2846 + 1000 = 3846
Painting	–	2	500 assemblies	3.00	500 × 3.0 = 1500	1500	3846 + 1500 = 5346
Packing	–	2	500 assemblies	–	–	250	5346 + 250 = 5596

Total Production time = 5596 min = 93 hrs 16 min

unfinished goods that could not be sold. The result would be dissatisfied customers and no cash flow to pay the wages and suppliers' accounts.

Production management is concerned with preventing these sorts of situations arising. Let's see how this is done.

The production has to be organized so that the sequence of operations enables the work to flow through the factory in the most efficient and cost effective way possible. To do this a *production schedule* has to be produced. This is often in the form of a Gantt chart. These charts were introduced in Section 2.2.1. Let's now draw up a Gantt chart scheduling the production of our carrying boxes. In order to draw up the Gantt chart we have to use the operation times that we have just calculated in Section 3.1.5, and make the following assumptions:

- The factory operates a five day week (Monday to Friday inclusive).
- The daily hours are 08.00 to 12.30 hrs and 13.30 to 17.00 hrs = 8 hrs per day.
- Work movement (transit) times between operations are ignored for simplicity in this example.
- Critical control point inspection times are also ignored for simplicity. In any case they should be organized so as not to interrupt production if no faults are found that require correction.
- Sanding down and painting operations are relatively slow and can commence before all the boxes have been assembled.

Let's start our chart on Monday morning at 08.00 hrs. The first 2 minutes will be spent in setting up the circular saw to cut the blanks for the bases. Cutting the blanks only takes 125 minutes so it should be complete by 10.07 hrs. The machine can then be reset for the next operation, namely, the blanks for the ends. Assuming no snags and allowing for the 12.30 to 13.30 hrs break, the blanks should all be cut by 15.19 hrs.

Now comes the clever bit. The operations to finish the end panels (rounding the top edge, cutting the finger holes and slotting) are done on a different machine called a *router*. Providing it is available, these operations can be commenced immediately while the sawyer is cutting the side panels. However, life is never so convenient, so let's assume the router does not become available until 16.00 hrs on Monday. All routing operations will be completed by 11.50 hrs on Wednesday and, in the meantime, the side panels will be cut and waiting.

Assembly is a relatively slow operation so it can start once about half the final routing operations are complete, say 10.30 hrs on the Wednesday. Note that if two persons are carrying out the assembly, the actual time taken is $1500/2 = 750$ minutes. So if assembly commences at 10.30 hrs on Wednesday it will finish at 1600 hrs on Thursday.

Sanding down the rough edges and joints is quicker than assembly, so a pool of finished work needs to be built up before this operation can commence. It would be safe to start sanding down at 08.00 hrs on Thursday. Painting is a slower process than sanding so it can start almost straight away, say 09.00 hrs on Thursday.

Two operators perform the sanding process so the actual time taken will be $1000/2 = 500$ minutes. So if sanding starts on Thursday at 08.00 hrs, it will finish at 08.20 on Friday. Similarly painting can start at 09.00 hrs on Thursday and finish at 14.30 hrs on Friday.

The boxes would then be left to dry thoroughly over the weekend and packing would commence at 08.00 hrs on the following Monday. This would be completed by 12.10 hrs. Therefore the boxes could be loaded and ready for despatch immediately following the lunch break.

Figure 3.6 shows the finished Gantt chart scheduling the manufacture of our carrying boxes.

3.1.7 Producing a production plan

A production plan brings together all the factors discussed so far in the form of a table. Such a plan includes:

- identification of the product specification details including;
- a description of the key production stages and critical control points;
- identification of the resources required;
- the estimated processing times for each key production stage;
- a production schedule in the form of a Gantt chart.

Figure 3.7 shows the layout of a typical form for a production plan. The Gantt chart scheduling the production would then be derived from this form.

Test your knowledge 3.3

1. Draw up a Gantt chart scheduling a typical day in your life from the time you get up until you go to bed.

2. Draw up a production plan and schedule for a simple component of your own choosing.

Key Notes 3.1

- Production planning is essential to make certain that the machines and raw materials all become available at the correct time to ensure a product is manufactured to specification and delivered on time.
- A product specification summarizes all the information about a product that needs to be known for its manufacture.
- A production schedule sets out the time taken for each operation and relates them usually in the form of a Gantt chart so that the progress of manufacture can be seen at a glance.
- Critical control points are those points during the manufacturing process when it is essential to make quality checks before the work is passed forward for further processing or is sold to the customer.
- Quality indicators may be *attributes* that can only be right or wrong, or *variables* that can vary between specified limits.
- Capital resources are such things as buildings, plant and machinery.
- Tooling resources are consumable items such as cutting tools, jigs and fixtures.
- Material resources can be both raw materials from which the product is manufactured as well as items that are bought in finished and ready for incorporation into the product being manufactured, e.g. buttons for a shirt.
- Human resources refer to all the people involved, directly and indirectly, in the manufacture of products.
- Service resources refer to the gas, water, electricity, compressed air and process drainage required solely for manufacturing processes.

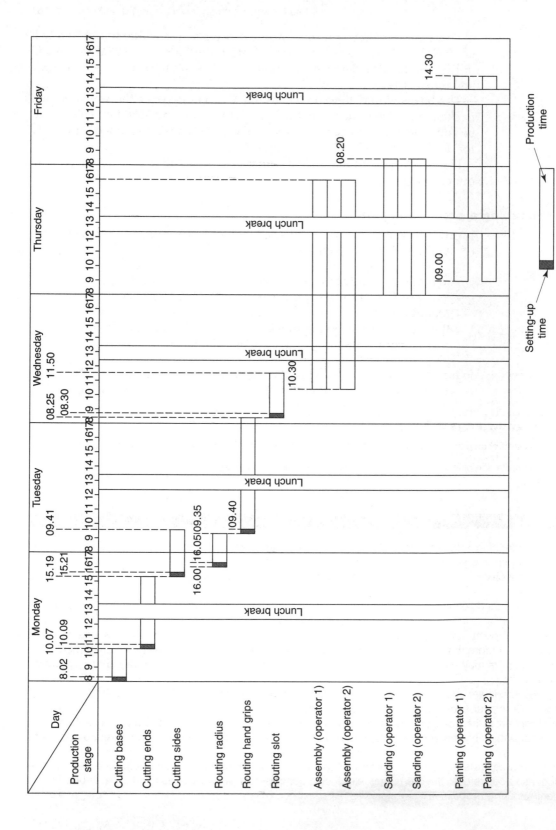

Figure 3.6 Gantt chart for scheduling the manufacture of the wooden carrying boxes.

Customer:		Product:		Design No.		Order No.		Quantity:		Date required:
Production stage	Capital resource	Human resource	Material resource	Service resource	Tooling resource	Machine set-up time	Number of operations	Operation time per-unit	Total operation time	Production stage time

									Total production time	
Notes:										

Figure 3.7 Production planning form.

• Processing time is the time taken to set up and perform a particular operation on either an individual product or a batch of products.

Evidence indicator activity 3.1

Draw up a production plan for a product of your own choosing, incorporating:

- a detailed product specification incorporating the critical control points;
- a detailed description of the key production stages;
- identification of the resources required;
- estimated processing times for each key production stage;
- a production schedule in the form of a Gantt chart.

3.2 Calculating the cost of a product

3.2.1 Direct costs

The cost of a product is made up of three main elements. These are:

- *Direct costs*, sometimes called *prime costs*, are such items as wages and materials that can be assigned directly to a particular product.
- *Indirect costs*, sometimes called *overhead costs* or *on-costs*, are those items that cannot be directly assigned to a particular product but have to be shared between all the different products manufactured by the company.
- *Profit*. This is superimposed on all the other costs and provides for such items as dividends, investment, reserve funds and taxation.

Therefore:

Total cost (selling price) = direct costs + indirect costs + profit

Let's now see how we can calculate the direct costs for a product. Again we can use a simple formula that simply consists of adding together the relevant cost elements for the resources used:

Direct costs = material resource costs + direct labour costs + service costs

Material resource costs

Material resource costs include *raw materials costs* such as:

- sheet metal, bars, forgings and casting in the engineering industry;
- rolls of cloth in the clothing industry;
- sawn timber, sheets of plywood and chipboard in the woodworking industries.

Material resource costs also include *component costs*, for example:

- individual components such nuts and bolts in the engineering industry, together with subassemblies such as water pumps and gear boxes in the automobile industry;

- buttons and zip fasteners in the clothing industry;
- nails, screws and adhesives in the woodworking industries.

Let's consider the material resource costs for our carrying boxes. We will require sheets of plywood, together with a supply of suitable adhesive and staples for power stapling guns. Hopefully the designer of the boxes took the standard sizes for sheets of plywood and chipboard into account in the initial design stage so that there will be as little waste as possible.

Note that chipboard and plywood sheets are sold in feet for the length and breadth and in millimetres for the thickness. Thus a standard 6 millimetre thick sheet of plywood will measure 4 foot by 8 foot.

Figure 3.8 shows how the components for our carrying boxes can be cut from the sheets so as to leave minimum waste. There is always some waste because the component dimensions must be largely governed by the items being carried in the boxes and also because of the width of each saw cut.

You can see from Fig. 3.8 that the following quantities can be cut from standard 4 ft by 8 ft sheets of plywood.

- *Base panels*. The wholesale price of 12 mm thick birch ply at the time of writing is £20.00 per sheet. Since 25 panels can be cut from a sheet, the cost of each panel is £20 ÷ 25 = **£0.80**.
- *End panels*. The wholesale price of 12 mm thick birch ply at the time of writing is £20.00 per sheet. Since 30 panels can be cut from a sheet, the cost of each panel is £20 ÷ 30 = **£0.67**.
- *Side panels*. The wholesale price of 6 mm thick birch ply at the time of writing is £11.85 per sheet. Since 15 panels can be cut from a sheet, the cost of each panel is £11.85 ÷ 15 = **£0.79**.

Each box consists of two side panels, two end panels and one base panel. Therefore the total timber cost for each box is:

$$(2 \times £0.67) + (2 \times £0.79) + (1 \times £0.80) = £1.34 + £1.58 + £0.80 = \textbf{£3.72}$$

In addition allowance must be made for the staples and glue used in assembly and a spray-on lacquer for the finish. These items would bring the total material resource cost up to **£3.90** per box.

Human resource costs

These costs will depend upon the quantity of boxes being produced. If only a small number (say 100) are being produced then they would most likely be made as follows:

- Cut the sheets of plywood into rectangles of the correct size on a circular saw.
- Radius the tops of the end panels on a bandsaw or with a portable power tool such as a jigsaw.
- Cut out the lifting slots in the end panel with a hand-held jigsaw.
- Cut the slots in the end panels with a portable router.
- Assemble by gluing and pinning.
- Paint.

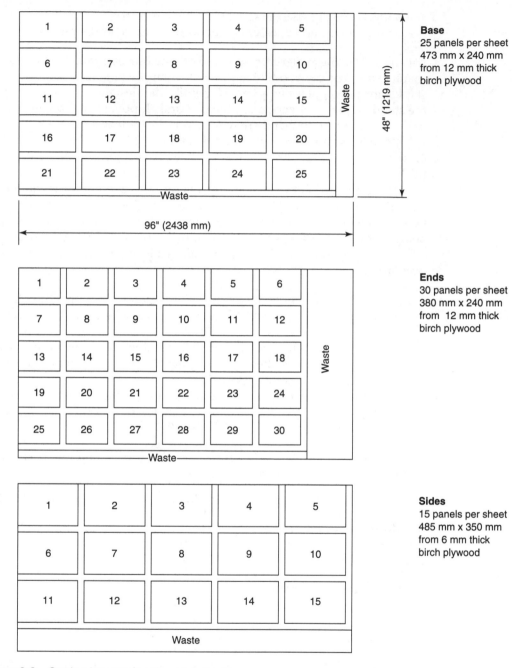

Figure 3.8 Cutting layouts for plywood sheets.

If a larger batch of boxes was being made (say 500) and if the order was likely to be repeated then it would be worthwhile making a set of steel templates for the router. These would allow the end panels to be profiled and for the lifting holes to be cut out on a router in addition to cutting the slot. This would not only be quicker, it would also be more accurate and save on hand finishing. It would also be worthwhile buying a power tacking machine for the assembly and also a hot glue gun. Painting would be by spraying rather than by brushing. In Section 3.2.3 we will look at break-even diagrams. These are used as an aid to determining when it is worthwhile investing in more productive equipment.

Let's now start to put some times and costs to direct labour involved in the making of our boxes. We can do this using a simple formula:

$$\text{Human resource cost} = \text{operation time} \times \text{hourly rate}$$

Remember to work in the same units throughout – for example, if the operation time works out to an answer in minutes or seconds, these must be converted to fractions of an hour if the cost element is based on an hourly rate.

Table 3.5 shows how the human resource cost is arrived at for 100 boxes and Table 3.6 shows how the human resource cost is arrived at for 500 boxes. Table 3.6 is based on the times arrived at in Table 3.4.

Table 3.5 Direct labour (human resources) costs for 100 boxes

Production stage	Production stage time	Hourly rate	Production cost
Cutting base blanks	30 min	£8.00	(30/60) × £8.00 = £4.00
Cutting end blanks	60 min	£8.00	(60/60) × £8.00 = £8.00
Cutting side blanks	60 min	£8.00	(60/60) × £8.00 = £8.00
Radiusing ends (bandsaw)	200 min	£8.00	(200/60) × £8.00 = £26.67
Cutting hand grips (jigsaw)	400 min	£8.00	(400/60) × £8.00 = £53.33
Cutting slots	60 min	£8.00	(60/60) × £8.00 = £8.00
Assembly, glueing and tacking	500 min	£5.00	(500/60) × £5.00 = £41.67
Sanding and dressing	400 min	£3.50	(400/60) × £3.50 = £23.33
Painting	400 min	£5.00	(400/60) × £5.00 = £33.33
Packing	60 min	£3.50	(60/60) × £3.50 = £3.50

Total human resource costs = £209.83
Unit cost per box = £2.098

* *Note*: Since the hourly rate is used, the production stage time in minutes had to be converted to hours by dividing by 60.

Services resource cost

As stated earlier in this chapter, this is the cost of the fuel gas, water and electricity used in the manufacture of the firm's products. Compressed air depends upon electricity to drive the compressor so it can be included with the electricity used directly in the processing. The same formula can be used no matter which service is used, but the units are different so some care is required:

$$\text{Services resource cost} = \text{cost per unit} \times \text{number of units used}$$

Electricity

The cost of electricity depends upon the number of *units* used. The unit is the kWh. This stands for kilowatt-hour. A kilowatt is 1000 watts. Therefore

Table 3.6 Direct labour (human resource) costs for 500 boxes

Production stage	Production stage time	Hourly rate	Production cost
Cutting base blanks	127 min	£8.00	$(127/60) \times £8.00 = £16.93$
Cutting end blanks	252 min	£8.00	$(252/60) \times £8.00 = £33.60$
Cutting side blanks	202 min	£8.00	$(202/60) \times £8.00 = £26.93$
Radiusing ends	155 min	£8.00	$(155/60) \times £8.00 = £20.67$
Cutting hand grips	405 min	£8.00	$(405/60) \times £8.00 = £54.00$
Cutting slots	205 min	£8.00	$(205/60) \times £8.00 = £27.33$
Assembly and glueing	1500 min	£5.00	$(1500/60) \times £5.00 = £125.00$
Sanding	1000 min	£3.50	$(1000/60) \times £3.50 = £58.33$
Painting	1500 min	£5.00	$(1500/60) \times £5.00 = £125.00$
Packing	250 min	£3.50	$(250/60) \times £3.50 = £14.58$
			Total human resource costs = £502.37

Notes: 1. Since the hourly rate is used, the production stage time in minutes had to be converted to hours by dividing by 60.
2. Using greater mechanization when producing 500 boxes per batch reduced the cost per box (unit cost) from £2.098 to £1.005.

1 kWh equals the use of 1000 watts for 1 hour, or 1 watt for 1000 hours, or 250 watts for 4 hours, or any other combination of watts and hours that come to 1000 when multiplied together.

The cost per 'unit' for a domestic house at the time of writing is 7.18p in the West Midlands region of the UK. Therefore the cost of electricity can be calculated as follows:

$$\text{Cost of electricity} = \text{cost per unit} \times \text{number of units used}$$

Table 3.7 shows some examples of the number of 'units' of electricity used by the various appliances in a house in the course of a day and the resultant cost.

Table 3.7 Typical daily consumption of electricity for one household

Appliance	Rating (kW)	Time (hrs)	Energy consumption (kWh)
Lighting	0.8	4	$0.8\,kW \times 4\,hrs = 3.2\,kWh$
Heating	3.0	5	$3.0\,kW \times 5\,hrs = 15\,kWh$
Heating	2.0	3	$2.0\,kW \times 3\,hrs = 6\,kWh$
Immersion heater	3.0	2	$3.0\,kW \times 2\,hrs = 6\,kWh$
Cooker and hob	5.0	$1\frac{1}{2}$	$5.0\,kW \times 1.5\,hrs = 7.5\,kWh$
Electric kettle	2.0	$1\frac{1}{4}$	$2.0\,kW \times 1.25\,hrs = 2.5\,kWh$
Washing machine	3.0	$1\frac{1}{2}$	$3.0\,kW \times 1.5\,hrs = 4.5\,kWh$
Security light	1.0	5 hrs	$1.0\,kW \times 5\,hrs = 5.0\,kWh$

Total energy consumption per day = 49.7 kWh
Total cost per day = 49.7 kWh × 7.18p
= 356.85p (approx.)
= **£3.57p (approx.)**

The cost of electricity for manufactured goods can be calculated in exactly the same way except that the number of units used will usually be very much greater and the cost per unit will be charged at a different rate for bulk industrial users.

Fuel gas

Gas taken from the mains is charged for in a rather complicated way.

- The volume of gas consumed is registered on a gas meter in cubic metres.
- The 'quality' of the gas supplied can vary so you have to take into account its *calorific value* as stated on the bill. This is the heating effect of the gas. It is given in units of megajoules per cubic metre (MJ/m^3).
- The volume of gas used is multiplied by its calorific value and divided by 3.6 to give the gas used in kWh so you can compare the cost with that of electricity.

Fortunately all these sums are done for you by the gas board and are printed out on your bill. Note that while electricity costs 7.18p per kWh, gas costs 1.48p per kWh. Therefore gas fired central heating is very much cheaper than using electric radiators.

Again, the cost of using gas is calculated as in the previous example. If you use 5 kWh of gas then you will be charged as follows:

$$Cost of gas = cost per unit \times number of units used$$

$$= 1.48p \times 5$$

$$= \textbf{7.4p}$$

For some processes, such as welding and brazing, bottled gas is used – for example, propane, acetylene, oxygen, etc. These gases are charged for by the volume of the bottle (cylinder) in which they are stored and there is a deposit charge on the bottle (cylinder). Therefore the process cost will depend upon the number of bottles or cylinders used.

Water

Nowadays, water for many houses and all commercial and industrial premises is metered. It is charged for by the cubic metre and the cost varies widely from one water board to another. An average cost to a domestic consumer is about 70p per cubic metre. Remember that the water board also has a responsibility for removing waste water and sewage from your premises and for treating it to make it safe. Therefore rather more than half the cost of the water supplied is to cover the cost of waste removal and treatment.

Again, we can calculate the cost of water used in the same way. Suppose we use 5 m^3 of water over a 7 day period of time and water costs 70p per m^3. The cost per day is:

$$Cost of water for 7 days = cost per unit \times number of units used$$

$$= 70p \times 5 m^3$$

$$= £3.50$$

$$Therefore the cost per day = £3.50 \div 7$$

$$= \textbf{50p} per day$$

You may wonder why, in a book on manufacturing, domestic situations have been included. This is so a scale can be used with which you are familiar and also because, in the small firm situations used in the examples, the costs will be very similar. Large firms such as car plants and chemical plants are not only bulk users they also have special disposal problems. Therefore they negotiate directly with their resource suppliers for special prices. To use such examples would complicate the issues unnecessarily. Let's now see how the services resource costs can be applied to our carrying boxes. Then we can obtain the overall direct costs. Table 3.8 shows the service resource costs.

Table 3.8 Service costs for production of 500 boxes

Production stage	Machine power	Production stage time	Energy cost
Cutting base blanks	3 kW	127 min	$(127/60) \times 3\,kW \times 7.18p = 45.59p$
Cutting end blanks	3 kW	252 min	$(252/60) \times 3\,kW \times 7.18p = 90.47p$
Cutting side blanks	3 kW	202 min	$(202/60) \times 3\,kW \times 7.18p = 72.52p$
Routing end radii	2 kW	155 min	$(155/60) \times 2\,kW \times 7.18p = 37.10p$
Cutting hand grips	2 kW	405 min	$(405/60) \times 2\,kW \times 7.18p = 96.93p$
Cutting base slots	1.5 kW	205 min	$(205/60) \times 1.5\,kW \times 7.18p = 36.80p$
Glueing (heat gun)	0.25 kW	1500 min	$(1500/60) \times 0.25\,kW \times 7.18p = 44.88p$
Tacking (staple gun)	0.50 kW	250 min	$(250/60) \times 0.50\,kW \times 7.18p = 14.96p$
Painting (compressor)	2.5 kW	1500 min	$(1500/60) \times 2.5\,kW \times 7.18p = 448.75p$
			Total service cost = 888.0p
			= £8.88

Notes: 1. Only electrical services are required for this product.
2. The staple gun only uses electricity each time a staple is fired. It is not running continuously.
3. Electric motors only draw power from the mains in proportion to the load imposed upon them. In this example an average, constant value is given for simplicity.
4. The times are based on Table 3.4.
5. Since electrical energy is bought by the 'unit', and one 'unit' is 1 kWh, the times in minutes need to be converted to hours by dividing by 60.

From our three tables we can now arrive at the overall direct costs for making 500 boxes using the formula introduced at the start of this section:

$$\text{Direct costs} = \text{material resource} + \text{direct labour costs} + \text{service costs}$$

$$= (£3.90 \times 500) + £502.37 + £8.88$$

$$= £1950 + 502.37 + £8.88$$

$$= \textbf{£2461.25} \text{ for 500 boxes} \quad [\text{Unit cost £4.92 per box}]$$

3.2.2 Indirect costs

These are the costs that are incurred by a firm through its very existence. All the direct costs considered earlier where *variable costs*. They varied in proportion to the number of boxes made – for example, twice as many boxes would mean buying twice as much plywood. However, indirect costs may be divided into *fixed costs* and *variable costs*:

- *Fixed costs* such as rent, business rates, insurance, bank loan interest, maintenance costs, management salaries, etc., will have to be paid even when the firm is closed down for its annual holiday and no manufacturing is taking place.
- *Variable costs* will change with the level of business activity – for example, telephone and fax charges, stationery, postal costs, marketing, etc. The more active the company becomes the greater will be these charges. The relationship between fixed and variable costs are shown in Fig. 3.9. The total indirect costs are:

Fixed indirect costs + variable indirect costs for a given level of business activity

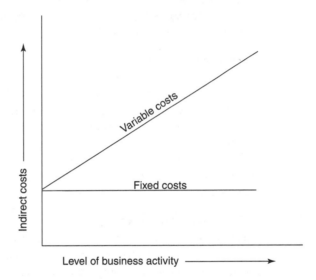

Figure 3.9 Fixed and variable costs.

Indirect costs vary widely from industry to industry and from firm to firm – for example, research and development costs in the pharmaceutical industry represent a major component in the selling prices of its products. In industries manufacturing relatively 'low-tech' products, the research and development charges are very much less. Let's now look at some representative indirect costs and see how they can be allocated to manufactured products.

Research and development costs

This can cover two aspects of manufacturing activity:

- the product being made;
- the method of manufacture.

All manufacturers must be constantly examining the market into which they are selling in order to find out:

- the needs of their customers;
- how their competitors are responding to these needs in an attempt to win business.

Product research and development must be a continuing process so that new and improved designs are always available to satisfy changes in the demands of the market. In order to keep ahead of its competitors, a successful firm tries to lead the market rather than follow it.

In addition, manufacturers must be constantly aware of changes in production technology and materials that can be adopted to make a better product more quickly and cheaply. The research and development department must be constantly assessing new developments in production technology and materials in order to advise the management when changes need to be made to keep the company at the forefront of the industry. It is then up to the senior management to assess whether the benefits from the improved technology warrant the expenditure involved.

Management costs

These involve management salaries and their secretarial support salaries. They also involve the cost of the so-called 'perks' such as a company car and health insurance. Many managers have to travel on a global basis in the conduct of the company's business and their travel and subsistence expenses must also be taken into account. Many managers also have to attend conferences and symposiums on new techniques and new national and international legislation and regulations. Conference fees are increasingly expensive.

Administration costs

These costs include the wages and salaries of clerks, typists and general office staff, together with telephone and postal charges, facsimile and photocopying charges, and stationery. They also include the cost of maintenance personnel, caretakers and security staff and cleaners. Nowadays, these latter costs tend to be 'outsourced' (contracted out) to specialist firms.

Marketing costs

These costs include advertising and product promotion, sales and marketing staff salaries and commissions. Again, market research, advertising and product promotion is increasingly 'outsourced' to specialist firms.

General expenses

These include rent, business rates, bank charges and interest, heating, lighting, depreciation, insurance, training and carriage.

Allocation of indirect costs

As already stated, indirect costs are those costs that cannot be associated directly with a particular product. Therefore it is difficult to allocate such costs to the selling price of the product. Various methods of allocating these costs are used. Some are relatively simple and some are extremely sophisticated. We are only concerned with two of the more simple possibilities.

Method 1

The direct human resource costs for a particular company come to a total of £100 000 per year. The indirect costs come to a total of £75 000 per year. Therefore the ratio of the indirect costs to the direct human resource costs is £75 000 ÷ £100 000 = 0.75. This ratio is called the *scaling factor*. We have already seen that:

$$\text{Total cost (selling price)} = \text{direct costs} + \text{indirect costs} + \text{profit}$$

The direct costs (*prime costs*) are the sum of the direct human resource costs plus the direct material costs plus the service costs.

The indirect costs are the direct human resource costs × 0.75.

For example, if the human resource costs for a particular product are £10 000, the material costs are £450 and the service resource costs are £50, then we can build up the selling price as follows:

$$\text{Total cost} = £10\,000 + £450 + £50 + (£10\,000 \times 0.75) + \text{profit}$$

$$= £10\,500 + £7500 + \text{profit}$$

$$= \textbf{£18\,375} + \textbf{profit}$$

The profit margin will depend upon a number of factors but must enable the company to put aside sufficient money to cover taxation, plant replacement, dividends and to build up a reserve fund against unforeseen emergencies.

Method 2

Where the material costs are relatively small compared with the direct human resource costs, many firms apply their scaling factor to the *total prime cost*. They argue that all materials have to be handled in and out of the stores, and that the rent, rates, heating and lighting that are required in servicing the stores area as well as the wages for the stores personnel all have to be paid for. The prime cost also includes the service resource costs but, as this is relatively small, it will be left in the prime cost in this example.

Further, the working capital (money) tied up in material stocks is losing the interest that it could earn if invested. Worse, the money to pay for the materials in the stores may have to be borrowed from the bank and loan charges will have to be paid to the bank.

Let's now look at the previous example again. The direct human resource costs for a particular company come to a total of £100 000 per year. The indirect costs come to a total of £75 000 per year. Therefore the ratio of the indirect costs to the direct human resource costs (scaling factor) is again £75 000 ÷ £100 000 = 0.75. We have already seen that:

$$\text{Total cost (selling price)} = \text{direct costs} + \text{indirect costs} + \text{profit}$$

The direct cost is the sum of the direct human resource costs plus the direct material costs plus the service resource costs. As previously, the human resource cost for a particular product is £10 000 the material cost is £450 and the service cost is £50. However, this time we will apply our scaling factor to the *total prime costs* instead of just to the direct labour costs.

We can now build up the selling price as follows:

$$\text{Total cost} = £10\,000 + £450 + £50 + (£10\,500 \times 0.75) + \text{profit}$$

$$= £10\,500 + £7875 + \text{profit}$$

$$= \textbf{£18\,375} + \textbf{profit}$$

Therefore we have gained £375 towards the cost of storing and handling the raw materials.

The profit margin will depend upon a number of factors. However, as stated previously, it must enable the company to put aside sufficient funds to cover taxation, plant replacement, dividends and to build up a reserve against unforeseen emergencies.

In some instances method 2 would inflate the price of the product unduly and make it uncompetitive. For example, to apply method 2 to the manufacture of jewellery using precious metal and precious stones would substantially increase the selling price compared with method 1. For many applications much more sophisticated methods for apportioning the indirect costs have to be used.

Using method 1, let's now add the indirect costs to the direct (prime) costs we have already worked out for manufacturing 500 of our wooden carrying boxes, using a scaling factor of 0.75.

$$\text{Total cost} = \text{direct costs} + \text{indirect costs} + \text{profit}$$

$$= £2461.25 + (£502.37 \times 0.75) + \text{profit}$$

$$= £2461.25 + £377.05 + \text{profit}$$

$$= \textbf{£2838.30} + \textbf{profit} \quad [\text{Unit cost £5.68 each}]$$

Test your knowledge 3.3

1. Explain the difference between direct and indirect costs.

2. Calculate the total cost of manufacturing 750 of our carrying boxes using chipboard at £6.50 per sheet and applying the indirect costs by method 2.

3.2.3 Effect of changing the scale of production

The effects of changing the scale of production were considered in Section 1.2.2. Generally, increasing the scale of production decreases the manufacturing costs since it is economical to increase the level of mechanization and automation. This was shown in a simple way when we compared the manufacturing costs for making our boxes in batches of 100, with the manufacturing costs when making our boxes in batches of 500 using improved production techniques.

However, we have to take into account the cost of purchasing and installing more productive equipment. To help management decide at what point it is cost effective to invest in improved production equipment, *break-even* diagrams are used. Let's look at a simple example.

To mark out, centre punch and drill a hole in a certain component costs £1.00 per hole. To drill the same hole using a drill jig reduces the cost to £0.50 per hole. The cost of the jig is £200. We need to know how many holes we have to drill for it to become cost effective to purchase the jig.

Figure 3.10 shows a break-even diagram for this situation. The line AB represents the cost of drilling the holes without the use of a jig. The cost is proportional to the number of holes drilled. Two hundred holes cost £200, 400 holes cost £400 and so on. The line CD is the cost of the jig. This is a fixed cost of £200 no matter how many holes are drilled. The line CE is the variable cost of drilling the holes using the drill jig. The cost of drilling the holes is added on to the cost of the jig.

Cost of jig drilling 200 holes = (200 × £0.50) + £200 for the jig = £300.

- Where the lines AB and CF cross, the cost is the same for both methods. This is the *break-even point*. It represents the minimum number of holes where it is cost effective to use the jig.
- For fewer holes, the cost of the jig is not warranted. The cost of the jig would not be recovered.
- As the number of holes to be drilled increases beyond that at the break-even point, it becomes increasingly profitable to use the jig.

Test your knowledge 3.4

1. A battery electric screwdriver costs £120. To insert screws in wood panelling using a conventional screwdriver costs 50p per screw. To insert screws in wood panelling using the electric screwdriver costs 10p per screw. Draw up a break-even diagram to determine the number of screws that must be inserted in the panelling for it to become cost effective to use the electric screwdriver.

Key Notes 3.2

- Direct costs (also called prime costs) are those costs that can be charged directly to a product, for example the material costs, the human resource costs and the service resource costs.
- Direct costs = material resource costs + human resource costs + service resource costs.
- Human resource costs = operation time × hourly rate.
- Service resource costs = cost per unit × number of units used.
- Indirect costs (also called overhead costs and on-costs) are those costs incurred in running the factory but which cannot be ascribed to any particular product, for example rent, rates, administrative staff salaries, telephones, maintenance, etc. These costs have to be apportioned between all the products manufactured and sold by the company.
- Profit is the difference between the sum of the direct and indirect costs and the final selling price, and provides for such items as dividends, investment, reserve funds and taxation.
- Fixed costs are those ongoing costs that are incurred even when no production is taking place, for example rent, rates, bank charges, management salaries, insurance premiums, etc.
- Variable costs are those costs that depend upon the level of business activity, for example telephone charges, postage, stationery, marketing, etc.
- Material resources consist of the raw materials required to manufacture a product. Material resources also include finished products such as nuts, bolts, buttons, zip fasteners, etc.
- Human resources consist of all the people required in the manufacture of a product. These can be divided into *direct labour*, such as fitters, operators, machinists, etc., and *indirect labour* such as designers, production engineers and supervisors, etc.

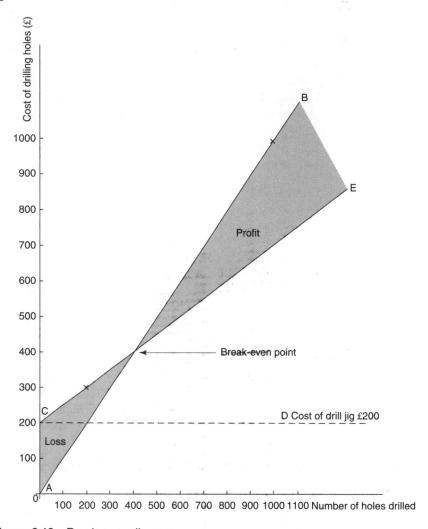

Figure 3.10 Break-even diagram.

- Service resources consist of such items as the gas, water, electricity and waste disposal that can be directly assigned to any given product. These are not to be confused with similar services required in the general running of the factory, for example lighting, heating, telephones, water for drinking and washing, etc.
- Total selling price = direct costs + indirect costs + profit.
- Break-even diagrams are graphs used to indicate the point when the volume of manufacture warrants investment in more productive plant.
- Break-even point is the point on a break-even diagram where the purchase of more productive plant becomes cost effective.

Evidence indicator activity 3.2

For a product of your own choice, record and calculate in detail the costs incurred during manufacture. These should include:

- the calculation of direct costs for each key stage of production;
- the estimation of indirect costs for each key stage of production;
- the calculation of the total cost of manufacture.

3.3.1 The key factors of a quality assurance system

Back in Section 1.2.5 quality was defined as 'fitness for purpose' and it was stressed that all manufactured products must satisfy the customer's requirements and represent value for money.

Quality assurance

Quality assurance is the result of creating and maintaining a quality management system that ensures all finished products achieve *fitness for purpose* and *conform* to the specifications agreed with the customer.

In the UK, ISO 9000 lays down guide rules for a quality management system that is recognized and approved internationally. To obtain the ISO 9000 kite mark a company must have its quality management system assessed by an independent, external certification organization. If it passes this *audit* and the company conforms to the requirements of the national standard, then its name appears in the Department of Trade and Industry (DTI) *Register of Quality Assessed United Kingdom Companies*. It can then trade with other similarly accredited companies throughout the world and its customers can be *assured* of the quality of its products.

To achieve these aims, a quality assurance system needs to be integrated with all the activities of a company. It needs to be integrated into the design and manufacturing processes of the company. It needs to be integrated with the sourcing and procurement of all materials used by the company and, finally, it needs to be integrated into the after-sales service. There must be total commitment to quality from the most senior management at boardroom level down to the newest trainee.

In order to build quality assurance into all stages and aspects of the manufacturing process, it is necessary to have a system of *quality control*.

Quality control

Quality control is the practical means of achieving quality assurance. It is the measuring of accuracy, the checking of finish and the maintenance of records, all of which are necessary to maintain the quality standards of production. To achieve quality control, quality standards must be agreed with the customer and the manufacturing processes must be planned so that these standards are achieved.

At key stages during the manufacture of the product (*critical control points*) it must be inspected and either accepted or rejected to ensure that quality standards are achieved. If accepted, the product is passed forward for further processing or sold on to the customer. If rejected, it must be scrapped or possibly recovered by reworking. Scrapping and reworking represent waste. Therefore most companies are now fully committed to a *right first time* policy.

As its name suggests a *right first time* policy depends upon the selection of manufacturing plant and processes that will ensure that reject components are kept at a minimum or eliminated altogether. Assemblies built up from quality assured components can then be expected to function correctly. Quality cannot be 'inspected into' a product. The product has to be made correctly from the start so that it can be passed by the inspection process. It has to be *right first time*.

Now let's consider the key factors of a quality control system that are essential to achieve quality assurance. These key factors can be summarized as:

- organization of the workforce;
- control of design;
- control of production systems;
- manufacture to specification;
- standards.

Let's now consider these key factors in more detail.

Organization of the workforce

At one time, quality was solely the responsibility of the chief inspector and his/her staff. Rejected work meant loss of bonus payments and led to industrial disputes. However important, the inspector was not a popular person.

Nowadays, quality is everybody's business from the managing director down. Staff training and organization are geared closely to a 'right first time' philosophy of quality. Through team meetings, project groups and 'quality circles', everyone is invited to participate in ensuring the quality of the company's products. At the same time, such meetings are also useful in developing working methods to improve productivity. By allowing the workforce to participate in problem solving, many of the human relation difficulties resulting from imposed solutions are removed.

To achieve these aims it is important that the company has a clearly defined organization chart (see Section 1.3.2) so that everyone knows what their responsibilities are and to whom they are responsible for quality. This is particularly important for supervisors and managers. Equally important is the encouragement of a closely knit team spirit, and many companies make use of outdoor pursuit centres to develop the leadership and collaborative qualities of their staff in 'bonding' exercises.

Control of design

As at every other stage of manufacture, quality should be 'built in' at the design stage. The factors that must be taken into account by the design team are:

- compliance with the customer's specification;
- compliance with the appropriate national and international regulations and codes of practice for the product being designed;
- Ensuring that the design can be manufactured to the quality standards required;

- the utilization of standard components manufactured to accepted standards wherever possible.

Control of the production system

The production system adopted for any given product will depend largely upon the scale of production.

Small scale

The product will be made within a jobbing workshop using general purpose equipment. The layout of a typical jobbing workshop was shown in Fig. 1.32.

Large scale

The product will be made in a workshop specially planned for the purpose. The machinery will have been selected to ensure that:

- the required rates of production can be achieved;
- the quality can be maintained as far as possible on a right first time basis;
- the cost of manufacture is competitive.

The workshop layout will have been planned to ensure that the materials and work in progress can be moved easily between the machines. Small products can be handled in storage bins on platform trucks. Larger products may need to be handled on pallets using forklift trucks. The largest products may require the availability of overhead cranes.

In both examples, production planning must ensure the availability of suitable raw materials at competitive prices. It must also allow for the handling and – if possible – reclaiming of substandard parts.

Standards

A *specification* can be defined as a document that lists the requirements with which a product has to conform – for example, a specification may be issued by a customer in order to ensure that the performance of a product meets with that customer's requirements.

Specifications are also drawn up by a national and international organization to ensure standardization, interchangeability and unity of quality (BS 7373 Guide to the Preparation of Specifications 1998). To manufacture to specification, a product must be inspected at various stages of its manufacture. The three bodies that are very important to the manufacturing industries are:

- The British Standard Institute (BSI)
- The European Norme (EN)
- The International Standards Organization (ISO)

To improve international trading all national standards organizations (such as BSI) are steadily harmonizing their individual standards by working with and adopting the recommendations of the ISO. You have already been introduced to ISO 9000 (formerly BS 5750).

ISO 9000 Quality Assurance (1987) was originally developed for the engineering sector as a means of developing a quality system for manufactures. ISO 9000 was developed from the former standard BS 5750 (1979).

Following on from the successful development and adoption of BS 5750 in the United Kingdom in 1987 it was adopted internationally as ISO 9000 and in 1992 further adopted by the European Community as EN 29000. The full quality assurance standard number is BS/EN/ISO 9000 abbreviated to ISO 9000.

The structure of the standard is as follows:

ISO 9000 (formerly BS 5750 part 0)

Guide to Management and Quality System Elements and is a guide to management responsibility, policy and objectives in implementing a documented quality system.

ISO 9000 (part 3) Guide to the application and management of software.

ISO 9001 (formerly BS 5750 part 1 1979) Specification for design, development, installation and servicing

The standard specification ISO 9001 is concerned with the system to be applied when the technical material requirements are specified in terms of performance or where a design has not yet been established.

ISO 9002 (formerly BS 5750 part 2 1979) Specification for development, installation and servicing

This standard is concerned with the system to be applied when the technical material requirements are specified in terms of an established design, and where specified product requirements are inspected and tested during manufacture.

ISO 9003 (formerly BS 5750 part 3 1979) Specification for final inspection and test

This standard is concerned with the system to be applied when the specified product requirements can be inspected and tested at the final stage of manufacture.

ISO 9004 (formerly BS 5750 part 4 1979) Quality management elements

ISO 9004 part 1 Guide to quality management and quality systems elements for services. Typically, the quality requirements cover such areas as:

- establishing and maintaining the system
- management responsibility
- design and document control
- contract review
- sourcing and purchasing
- process control
- control of bought out items and products
- control of inspection, measuring and testing equipment
- corrective actions
- quality records
- personnel and training.

Another important standard that is now having an important impact on manufacturing is ISO 14001 Environmental Management Systems (1996) and covers the following areas:

- Organization
- Environmental policy
- Planning – Legal requirements
- Training, documentation, emergency response
- Monitoring and measurement.

Test your knowledge 3.5

1. Briefly describe what is meant by the term 'quality assurance' and explain how it can be achieved.

2. Briefly describe what is meant by the term 'quality control'.

3. List the main advantages of a 'right first time' policy.

4. Describe the purpose of ISO 9000.

3.3.2 Function of quality indicators and critical control points

Quality indicators

Quality indicators and critical control points were introduced earlier in this book in Section 1.2.5. Remember that quality indicators can be classed as *variables* or *attributes*.

Variables

Dimensions are an example of a variable attribute since they cannot be manufactured nor measured to an exact size. Allowance must be made for some variation – for example, a dimension may be specified as:

$$35.58 \pm 0.08 \, mm$$

This means that the dimension can be considered as correct if it lies between:

$$35.58 - 0.08 \, mm = \textbf{35.50 mm} \text{ and}$$

$$35.58 + 0.08 \, mm = \textbf{36.60 mm} \text{ inclusive}$$

These are called *the limits of size* for this dimension. If the dimension lies *within limits* it is correct and will pass inspection.

The numerical difference between the limits of size is called the *tolerance*. In this example it is $36.60 \, mm - 35.50 \, mm = \textbf{0.10 mm}$. This is the amount of *deviation* in size the design will *tolerate* and still function correctly and meet its quality specification.

Attributes

These are quality indicators that do not allow for variation. They are either right or wrong – for example, steel oil drums will either leak or they will not leak. If they *leak* they are useless and are *rejected*. If they *do not leak* they will do the job for which they are intended and are *accepted*.

Critical control points

Critical control points are the points in the manufacturing sequence for a product where the product is checked to insure that it is being correctly manufactured and is to specification. We have already identified several critical control points in previous examples. These control points start with the raw materials and bought-in components, and continue through key stages of production, until the finished product is given its final inspection before release to the customer. *Quality indicators* are applied at each *control point*.

Test your knowledge 3.6

1. Describe, with examples, what is meant by the terms 'variables' and 'attributes' as applied to quality control.

2. For a product of your own choosing specify the quality indicators and draw up a production flow chart showing the positions of the critical control points.

3.3.3 The role of testing and comparison in quality control

Testing and comparison is essential in quality control in order to establish whether a product conforms to its specification. Let's see what this means.

We can test a product to determine such *variables* as its length, weight, volume or performance and compare the test results with the specification for the product. Provided the variables lie within the limits set by the designer the product is accepted. It is said to *conform to its specification*.

The product can also be inspected visually for such *attributes* as colour, finish, texture and/or fluid tightness. Provided such attributes are acceptable, then the product is said to *conform to its specification*.

3.3.4 Quality control techniques

If single components or small batches of components are being inspected then all the components will be examined. This is 100% inspection. Usually 100% inspection is too expensive where large quantities of components are being made such as nuts and bolts. Here, it would be normal to use a sampling technique. The exception would be for critical components used in an aircraft where many lives are at stake. In this instance, the safety of human lives is much more important than the cost of 100% inspection. This is one reason why aircraft are so costly. Before we consider sampling techniques we need to look at the various inspection and testing techniques used to maintain quality control.

Non-destructive testing

These are testing techniques that do not destroy or damage the product or the material sample being tested. This is essential if 100% inspection is being carried out.

Visual inspection

The colour, finish and texture of a manufactured product can be inspected visually with or without the aid of a magnifying glass or a microscope. Various charts are available for comparison. Surface cracks in castings can be identified by visual inspection and the superficial quality of welds can also be checked. Feel is also associated with visual inspection and surface finish can be checked against sample surfaces. However, it must be remembered that the human senses are comparative and not absolute. You can tell if one object is warmer than another object but you cannot say what its exact temperature is. All inspection techniques relying on the human senses are highly subjective and inherently unreliable as they depend upon individual sensitivity and levels of experience.

Dye penetrants

These are used to check for surface cracks and porosity in castings and forgings that cannot be detected by the human eye unaided. The penetrant fluid is painted onto the surface of the casting. After waiting a short while for the fluid to be absorbed into the surface defect, the surplus penetrant is wiped off and a white powder is dusted over the surface. As the penetrant oozes out of the surface defect, it stains the white powder and indicates the position and extent of the defect. A variation on this test is to use a penetrant that will glow (fluoresce) when exposed to ultraviolet light.

Magnetic detection

When metal components that are made from iron and steel are placed in a magnetic field, the magnetic flux travels through the metal. If there is a flaw in the metal at or near the surface the magnetic flux distribution is distorted. This distortion is detected by a suspension of magnetic powder (magnetic iron oxide) in paraffin or light machine oil spread thinly over the surface of the metal. The magnetic powder 'bunches' in the vicinity of the flaw as shown in Fig. 3.11.

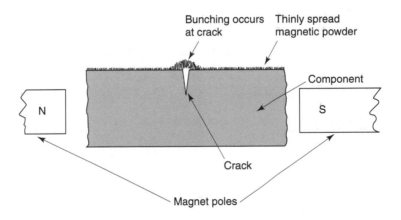

Figure 3.11 Principle of magnetic crack detection.

Ultrasonic testing

Ultrasonic testing (also called ultrasound) uses sound waves that are too high in frequency for the human ear to detect. Although it varies from person to person, on average the maximum frequency a person can hear is about 18 kHz. The frequencies used for testing range between 0.5 MHz and 15 MHz depending on the material being tested.

In ultrasonic testing, pulses of these high frequency oscillations are generated electronically and fed into the component being tested by a transducer.

The pulses are bounced back by any discontinuity in the material and will show up on a computer screen linked to the apparatus. The general principles are shown in Fig. 3.12. The advantage of ultrasonic testing is that it can be used with any material, magnetic or non-magnetic, metal or non-metal. It even works with human tissues and allows us to have painless and safe internal examinations without the need for intrusive surgery.

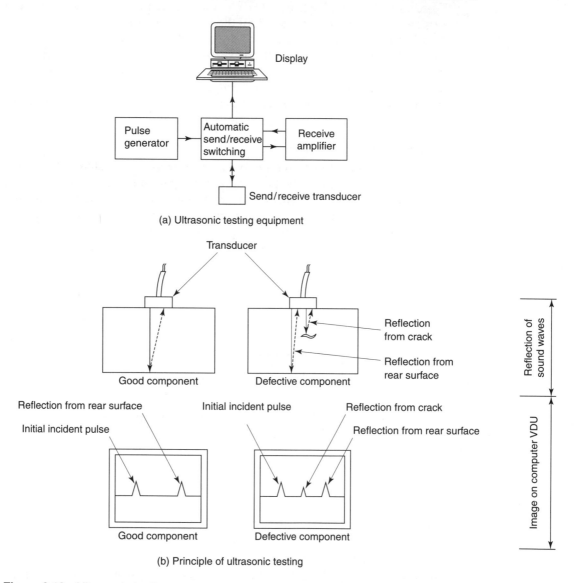

Figure 3.12 Ultrasonic testing.

Eddy current testing

When the varying magnetic field produced by an alternating current source is linked with a metal object, the magnetic field induces electric currents in that object. These electric currents are called *eddy currents*. This applies to all metals no matter whether they are magnetic or non-magnetic. The eddy currents themselves also produce magnetic fields of their own. The initiating magnetic field and the eddy current magnetic field react together to produce a resultant field that induces an electric current in a detector.

Providing the conditions remain constant as the detector is moved across the surface of the metal the resultant field and the electric current in the detector also remain constant. However, any discontinuity in the material causes the balance to be disturbed and the nature of the discontinuity can be displayed on a computer screen linked to the apparatus. The principles of this test are shown in Fig. 3.13.

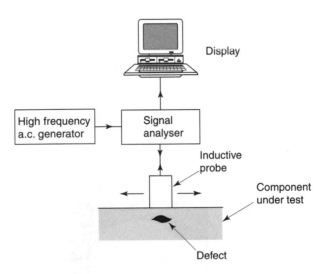

Figure 3.13 Principle of eddy-current testing.

Radiography

This is widely used for the examination of joints in welded pressure vessels such as boilers. Both X-rays and the more penetrating gamma rays are used depending on the type and thickness of the material being examined. Solid objects absorb the rays to a greater or lesser degree and cast a shadow on a photographic film sensitive to such rays as shown in Fig. 3.14(a). If there is a discontinuity fewer rays are absorbed and the discontinuity shows up as a darker area on the photographic film. This is shown in Fig. 3.14(b). The rays used in these tests are extremely dangerous and the most stringent safety precautions have to be taken. Radiographic tests must only be carried out by specially trained staff.

Electrical and electronic testing

Many of the tests described above use electronic and electrical equipment, and devices such as electronic scales are nowadays more likely to be used for weighing as they are more accurate. Testing devices based on the science of electronics have the advantage that they can be easily linked to plant control equipment and computers. When linked to computers, the test results can be:

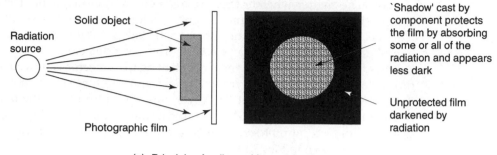

(a) Principle of radiographic examination

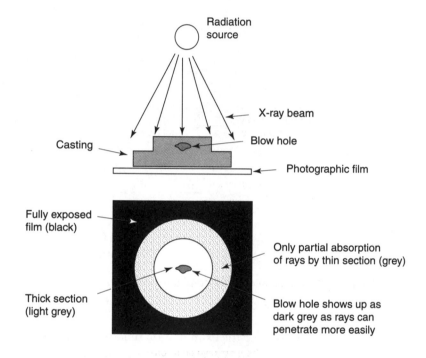

(b) How a defect is shown up by radiographic examination

Figure 3.14 Radiography.

- observed immediately on screen;
- printed out as hard copy;
- stored for future reference;
- subjected to statistical analysis.

However, electronic and electrical testing refers generally to the measurement of performance of electrical equipment. For example:

- the measurement of potential using voltmeters;
- the measurement of current flow using ammeters;
- the measurement of resistance and circuit continuity using ohmmeters;

- the measurement of amplifier performance (distortion and gain) using oscilloscopes.

Chemical analysis

Many substances have to be sampled to establish their purity and freedom from contamination. The samples are taken to a laboratory for chemical analysis. These include both the raw materials and the finished products. For example:

- food stuffs;
- pharmaceuticals;
- perfumes and toiletries;
- dyestuffs;
- fertilizers;
- drinking water;
- alcoholic and non-alcoholic drinks;
- liquid effluents before discharge into the sewers and water courses such as rivers.

Dimensional measurement

Dimensional measurement is relatively slow and expensive. It requires skill and experience in the use of the equipment, for example the use of micrometer and vernier calipers, the use of slip gauges and comparators, and the use of angular measuring equipment such as the sine bar. Such measuring equipment is mostly used when inspecting one-off and small batch production. For large batch and mass production, *gauges* are mostly used. These do not measure the component but merely indicate whether or not the dimension is within the limits laid down by the designer. Figure 3.15 shows some typical limit gauges. They all have GO and NOT GO elements. For example:

- If the GO end of a plug gauge enters a hole, then the hole could be the correct size. We shall only know if it is the correct size if the NOT GO end of the gauge will not enter the hole.
- If the GO end of the gauge will not enter the hole, then the hole is undersize and the component is rejected.
- If the NOT GO end of the gauge enters the hole then the hole is oversize and the component is rejected.

The use of limit gauges is much quicker and requires much less skill and training than the measurement of dimension. However, a gauge can only tell us if the dimension is within limits, it cannot tell the actual size.

Destructive testing

Most of the tests for determining the mechanical properties of materials are destructive tests. Therefore these are carried out on samples of the raw materials since it would not be practical to destroy the finished product. There are exceptions and, for crucial components and assemblies in the aircraft industry, sample wings and control surfaces, etc., are tested to destruction in

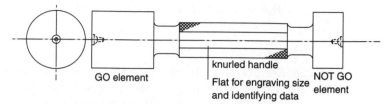
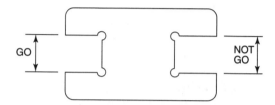

(a) Solid GO and NOT GO plug gauge for checking hole diameters

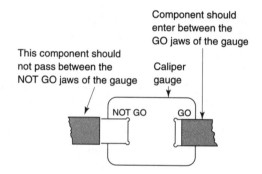

(b) Double ended GO and NOT GO caliper gauge for checking thicknesses

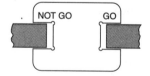

(a) Correctly sized component enters 'GO' jaws but not 'NOT GO' jaws

(b) Undersize component enters 'GO' and 'NOT GO'

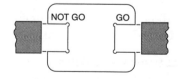

(c) Oversize component does not enter 'GO' or 'NOT GO'

(c) Use of caliper gauge

Figure 3.15 Use of gauges.

special test rigs to determine their strength and fatigue life. This enables the maintenance manual to indicate how frequently key components should be changed so that there is little or no danger of them failing in service.

Tensile test

Standard test specimens are subjected to tensile (stretching) loads. For each increase in load, the corresponding stretch is measured until the component eventually breaks. Figure 3.16 shows a typical testing machine, a typical specimen and the results of a test on a mild steel specimen expressed as a graph. Fibres, threads and cords used in such industries as fabric manufacture, rope making and tyre making are tested in this way as well as metals for the engineering industries.

Impact testing

This is used to determine the toughness of a metal. Care must be taken not to confuse strength and toughness. Strength is measured by the tensile test just described, whereas toughness relates to the impact strength of a material. Some brittle materials can have a high tensile strength but little resistance to impact.

Impact testing consists of striking a notched specimen of the material under test with a controlled blow. A typical test is the Charpy test and details of this and the testing machine are shown in Fig. 3.17.

Compression testing

This is the opposite of tensile testing, since the specimen is subjected to a compressive (squeezing) load. This is particularly important for materials such as building bricks, insulation blocks and concrete.

Hardness testing

Hardness is defined as the resistance of a material to indentation or scratching by another hard body. This is precisely how materials are tested for hardness. Various standard *indenters* are pressed into the surface of the specimen by a controlled load and the degree of penetration is measured. The form of the indenter varies with the type of test used – for example, the Brinell test uses a hard steel ball, the Vickers test uses a diamond pyramid and the Rockwell test uses a diamond cone. Soft materials such as elastomers (rubbers) and some plastics often require special tests. The principle and some test conditions for a Brinell test are shown in Fig. 3.18. The hardness of the specimen is determined by measuring the diameter 'd' and looking up the corresponding hardness number in the tables provided. Testing the hardness and abrasion resistance of materials that are going to be used for such purposes as ball and roller bearings, floor tiles and car windscreens is very important.

Fatigue testing

We have already seen that the testing of crucial aircraft components such as wings and control surfaces are often tested to destruction in special test rigs. In fact, this is *fatigue testing*. The rigs keep repeating the loads on the

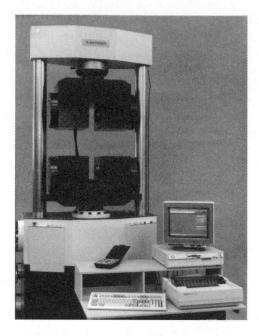

(a) Typical tensile testing machine. Courtesy of Samuel Dennison Ltd (Dennison Mays Group)

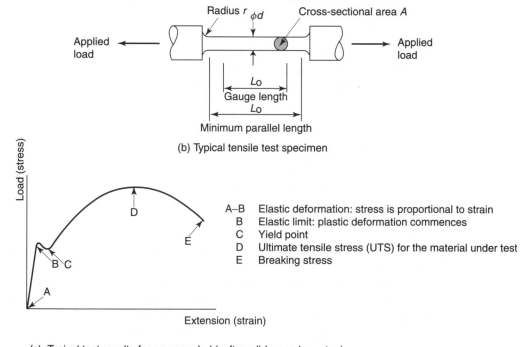

(b) Typical tensile test specimen

A–B	Elastic deformation: stress is proportional to strain
B	Elastic limit: plastic deformation commences
C	Yield point
D	Ultimate tensile stress (UTS) for the material under test
E	Breaking stress

(c) Typical test results for an annealed (softened) low carbon steel

Figure 3.16   Tensile test.

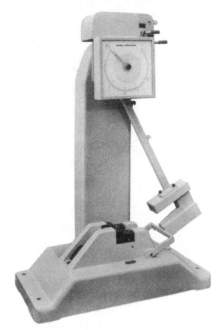

(a) Typical impact-testing machine.
Courtesy of Samuel Dennison Ltd (Dennison Mays Group)

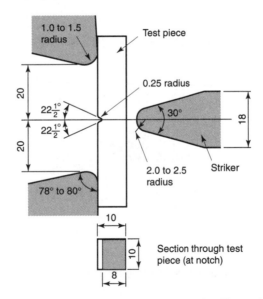

(b) Principle of the Charpy test (all dimensions in millimetres)

Figure 3.17 Impact testing.

components under simulated flight conditions until they fail. The equivalent of many years of flying can be reproduced in just a few hours in this way.

The principles of a simple fatigue test are shown in Fig. 3.19. Such a test produces an alternating load. It is often not possible to achieve fatigue failure in many materials in a reasonable period of time so the material is

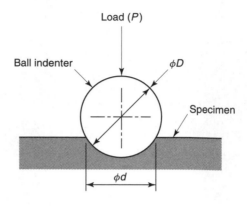

$$\frac{P}{D^2} = K$$

Where K is a constant; typical values of K are:

Ferrous metals	$K = 30$
Copper and copper alloys	$K = 10$
Aluminium and aluminium alloys	$K = 5$
Lead, tin and white-bearing metals	$K = 1$

ϕ d = diameter of indentation. Conversion tables are used to convert 'd' to a Brinell hardness number (H_B)

ϕ d is measured with a special microscope

Figure 3.18 Principle of the Brinell hardness test.

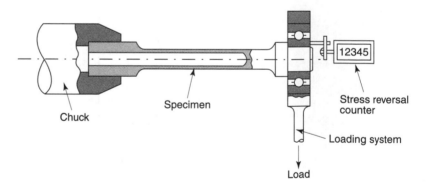

Figure 3.19 Simple fatigue test rig.

considered satisfactory if it has not failed by the time a prescribed number of stress reversals have been achieved. Fatigue testing is important for components ranging from such things as jet engine blades and motor car engine crankshafts, to springs and foam cushion fillings.

Expert scrutiny

This can be destructive or non-destructive. In destructive testing, sample tins of food, soft drinks, wine or other produce are laboratory tested for harmful bacteria. If safe, the scrutineer may then taste the produce to see if it is palatable and up to standard. This requires many years of experience and very expert knowledge. In non-destructive testing a scrutineer may examine coins, stamps and bank notes for production errors, or listen to musical instruments for quality of tone.

Test your knowledge 3.7

1. Describe the essential difference between strength and toughness.

2. For each of the following tests briefly describe a situation where the test would be most appropriate. Give the reason for your choice.

   (a) Ultra sonic testing.
   (b) Magnetic testing.
   (c) Radiographic testing.
   (d) Tensile test.
   (e) Charpy test.
   (f) Hardness test.

3. Explain why samples of key components for an aircraft are often tested to destruction in a special test rig that simulates flight conditions.

3.3.5 Sampling techniques

As previously stated, except for critical components upon which human lives depend, the cost of 100% inspection is too great for most manufactured goods. Therefore, where large batch or continuous production is involved, *sampling techniques* are used. Instead of inspecting all the parts in a batch, only a few samples are inspected. If the samples conform to the specified quality levels, then the whole batch is accepted. This is called acceptance sampling. If the samples do not conform then the whole batch is rejected or, sometimes, subjected to further inspection.

The clever part is in deciding how to sample the batch. If the sample is too large, the cost of inspection will be too great. If the sample is too small, the inspection results can be misleading. There are a number of sampling techniques. Let's now look at some of them.

Continuous sampling

This is the 100% inspection we have already considered for critical products. Every product is inspected as it is manufactured. This can only be afforded where human life is at stake, for example critical aircraft components, some foodstuffs and pharmaceuticals.

Random sampling

In random sampling, the first component of the batch is inspected and, if it conforms to the quality specification, production commences. From then on, samples are taken from the batch on a purely random basis for inspection. As with all statistical sampling techniques, the larger the sample the more accurate the results. If the sampling frequency and the size of the samples are reduced to economize on inspection costs, then sampling errors will occur.

Single sampling

This is a more controlled approach.

- A single batch of a prescribed size is taken.
- The number of defective components in the batch that can be accepted is also prescribed. This is called the *acceptance* figure.
- If the number of defective components in the batch is equal to or less than the acceptance figure then the batch is accepted. If greater, the batch is rejected.

The acceptance quality figure expressed as a percentage of the total batch size is called the Acceptable Quality Level (AQL) and must be agreed with the customer in advance of commencing production. A typical AQL is 5%. Let's see how this can be applied to a batch of 5000 components. The sample size is set at 80 components:

$$5\% \text{ of } 80 = 80 \times 5/100 = \textbf{4 components}$$

Therefore, if 4 or less of the components are defective then the whole batch is accepted. If more than 4 components are defective then the batch is rejected. So, if 5% defective components are an acceptable quality level, then 6% or more defective components are unacceptable to the customer. This unacceptable level of rejects is called the *Lot Tolerance Percentage Defective* (LPTD). So what happens if we find that, on inspecting the sample, there are slightly more than 5% defective components in the sample? We can:

- Inspect a second sample to confirm the findings of the first sample. If the second batch also contains more than 5% defective components then:
- Carry out 100% inspection of the batch to remove all rejects and rectify them if possible. This could be costly.
- Sell to a customer who will accept the higher AQL, possibly at a lower price (a discount).

Multiple sampling

Once we take more than a single sample we move into the more sophisticated technique of multiple sampling. There are three possible decisions that can be taken after inspecting a sample:

- Accept the whole batch.
- Reject the whole batch.
- Take further samples from the batch and add the results to the results of the first sample. An accept/reject decision is then taken on the overall result. This is usually done when the reject level is only marginally over the agreed AQL.

Test your knowledge 3.8

1. Explain where continuous sampling would be used and why it is expensive.

2. In your own words, state the essential differences between:

(a) random sampling;
(b) single sampling;
(c) multiple sampling.

3.3.6 Use of test and comparison data

We have already considered testing and comparison in Section 3.3.3. This involved comparing the test results with quality specifications and standards. Statistical techniques can be used to control our manufacturing processes and, through them, the quality of the manufactured products associated with those processes. The mathematical principles of statistics are beyond the scope of this book but the following simple example will show how such techniques can be used in practice.

By way of an example let's consider 5000 components that have to be machined to length from blanks sawn from bar material. The critical dimension requiring inspection is 75.00 ± 0.05 mm, that is, the acceptable size can vary from 74.95 mm to 75.05 mm inclusive. As in a previous example the single sample size is 80 and the AQL is 5%.

The sample is inspected and the results are shown in Table 3.9. In this example control is by variables so the lengths have been measured rather than gauged. The results are plotted graphically as a histogram as shown in Fig. 3.20. All the bars lie within the prescribed limits. Therefore the whole batch is acceptable.

Table 3.9 Sample lengths

Length (mm)	Number of bars
74.95 (Lower limit)	0
74.96	2
74.97	4
74.98	8
74.99	18
75.00 (Basic size)	22
75.01	17
75.02	6
75.03	2
75.04	1
75.05 (Upper limit)	0
Total	80

In practice an electronic comparator linked to a computer would most likely be used. The histogram would be shown on the screen and a hard copy printout made. The inspection results would be recorded permanently on disk for quality assurance purposes.

Repeat batches continue to be ordered and, inevitably, tool ware and resetting occur. Therefore each successive batch will vary slightly in the inspection results obtained.

Let's look at the table of inspection results (Table 3.10) and its associated histogram (Fig. 3.21) for the second batch sample. We can see that drift has

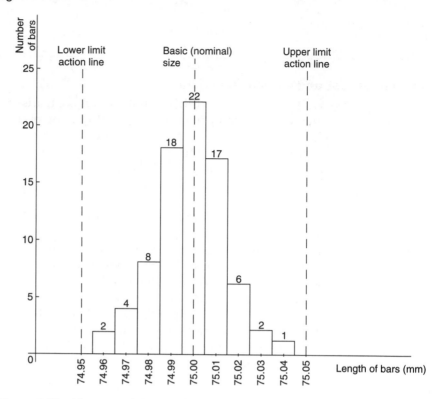

Figure 3.20 Histogram for the sample of 80 bars.

Table 3.10 Sample lengths showing drift

Length (mm)	Number of bars
74.95 (Lower limit)	0
74.96	0
74.97	0
74.98	0
74.99	2
75.00 (Basic size)	4
75.01	8
75.02	18
75.03	19
75.04	17
75.05 (Upper limit)	6
75.06	3
75.07	2
75.08	1
Total	80

occurred and some of the sample bars lie out of limits. They lie beyond the upper action line.

Since, in fact, 6 of the bars are over length out of a sample of 80 bars, this represents a percentage of $[6/80] \times 100 = 7.5\%$. This exceeds the AQL of 4% and the batch would be resampled and retested before being rejected

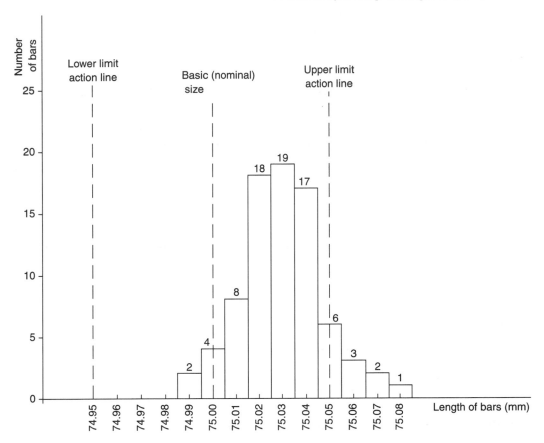

Figure 3.21   Histogram for the second sample of 80 bars.

or reworked. In theory, if the statistical method is sound, this result implies that out of the batch of 5000 bars 7.5% will be defective. There will be $5000 \times [7.5/100] = \mathbf{375}$ defective components.

Since the dimensional drift is towards the upper limit of size, the machine operator would be asked to reset the machine so that the inspected samples produce a histogram centred just below the 75 mm size. Let's now see how statistical process control can be applied in practice. There are two types of control chart that can be used:

- The *average control chart* where the average value of the variable being measured is calculated for each sample and is plotted on the control chart.
- The *range control chart* where the numerical difference between the greatest and least values for the variable being measured is calculated for each sample and is plotted on the control chart.

A typical average control chart is shown in Fig. 3.22. The *upper and lower action lines* are the dimensional limits stated on the drawing in conformance with the quality specification for the product. In addition there are *control* or

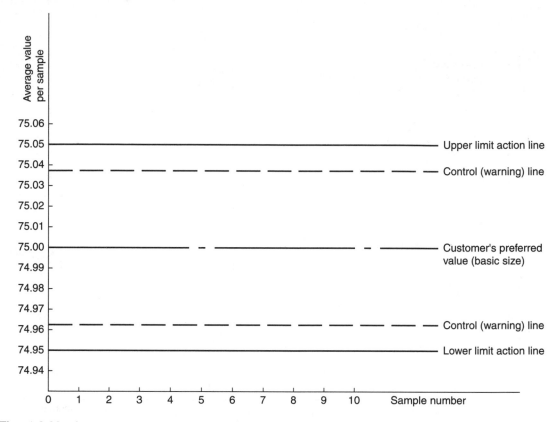

Figure 3.22 Average control chart layout for 75 mm long bars.

warning lines set just inside the action lines. These enable corrective action to be taken in advance of the action lines being exceeded and this reduces the chance of rejects produced and, worse, escaping the inspection net. The nominal value line is the customer's preferred value for the variable being measured. In our example this would be the 75 mm dimension.

The setting of the control lines is based on statistical calculations that are beyond the scope of this book. However, look at Fig. 3.23. This is a typical curve of normal distribution for a given population. In this context the word 'population' is the total batch from which the sample was taken. From the figure you can see that 68% of the population fall into a band spreading between +1 and −1 standard deviations. You can also see that 95% of the population will lie between +2 and −2 standard deviations. Finally almost 100% of the population (99.73% actual) will lie between +3 and −3 standard deviations. Don't worry about standard deviations, these can be obtained from statistical tables. What is important is how we use the charts. It is normal practice to set the control (warning) line at 2 standard deviations either side of the mean line. This means that 95% of our products should lie between them. It is also normal practice for the action lines to be set at 3 standard deviations. This should keep production well within the acceptable quality level (AQL) and give us *right first time* production.

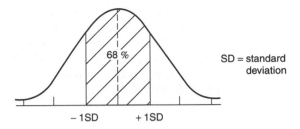

SD = standard
deviation

(a) 68% of all the total population will be contained under the curve
between +1 and –1 standard deviations about the mean value

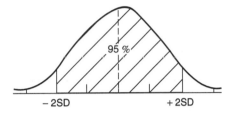

(b) 95% will be contained under the curve between +2 and –2
standard deviations about the mean value

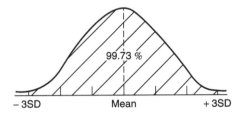

(c) Approximately 100% (99.73% actual) will be contained between
+3 and –3 standard deviations about the mean value, i.e. +3SD
gives approximately the upper limit of the distribution and –3SD
gives the lower limit of the distribution

Figure 3.23 Curve of normal distribution.

Let's now apply this to our example. The distribution of sizes for the
first inspection sample of boxes has already been listed in Table 3.9. We can
calculate the average value as follows:

$$A_v = \{(2 \times 74.96) + (4 \times 74.97) + (8 \times 74.98) + (18 \times 74.99) + (22 \times 75.00)$$

$$+ (17 \times 75.01) + (6 \times 75.02) + (2 \times 75.03) + (1 \times 75.04)\} \div 80$$

$$= \{149.92 + 299.88 + 599.84 + 1349.82 + 1650 + 1275.17$$

$$+ 450.12 + 150.06 + 75.04\} \div 80$$

$$= 5999.85 \div 80$$

$$= \textbf{744.998\,mm (75.00\,mm to 2 d.p.)}$$

We can also calculate the average value for our second Table (3.10) of
samples in the same way. Try it for yourself. You should obtain a value of
75.03 mm to 2 d.p.

Eight further batches of samples were inspected and the results obtained are as follows: 74.98 mm, 75.01 mm, 75.01 mm, 74.99 mm, 74.99 mm, 75.00 mm, 75.02 mm, 75.01 mm. All these results have been recorded on an average control chart of the type previously shown in Fig. 3.22. The completed chart is shown in Fig. 3.24. All the parts lie within the control lines so no action needs to be taken.

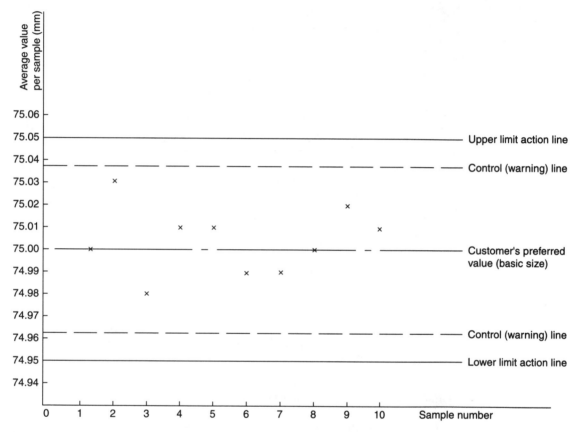

Figure 3.24 Average value control chart for 10 batches of 5000 bars each: inspection sample of 80 bars per batch.

Now we need to consider range charts. Figure 3.25 shows a typical range control chart. The chart shown is suitable for our 75 mm long bars. This time we are only showing the upper action line (upper action limit) and the upper control line (upper warning limit). The lower action and control lines are not shown, as any drift of component size towards them would only indicate *improved accuracy* and would be no cause for concern. For both types of control chart the limit lines are not calculated from first principles but are obtained from standard tables (see BS 2564). By setting the control limits at 2 standard errors, the probabilities are that only 2.5% of parts (1 in 40) will exceed this limit and corrective action can be taken before the AQL of 5% is exceeded. By setting the action line at 3 standard errors the chances are

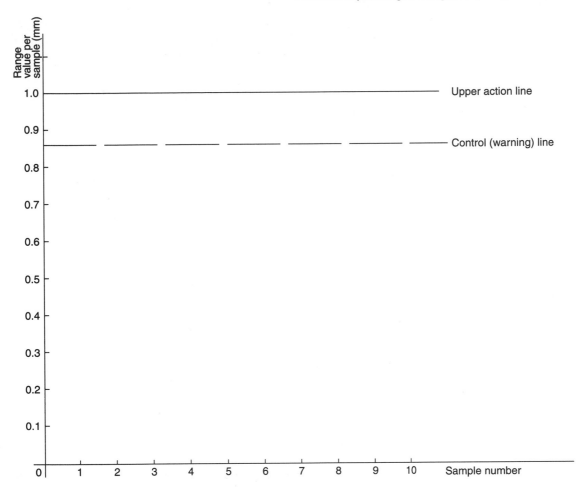

Figure 3.25 Range value control chart suitable for 75 mm long bars.

that only 0.1% (1 in 1000) will exceed this limit. Note that in this context the *standard error* is the standard deviation from the distribution of sample means. Don't worry, you won't have to calculate these values; they can be looked up in tables.

Let's now calculate the first range value from the data in Table 3.9.

$$R_v = \text{largest value} - \text{smallest value}$$

$$= 75.04 \, \text{mm} - 74.96 \, \text{mm}$$

$$= \mathbf{0.08 \, mm}$$

Similarly you can work out the range value for the data given in Table 3.10. Go on, try it for yourself. You should get a value of **0.09 mm**. The next eight batches gave samples having the following range values:

0.07 mm, 0.06 mm, 0.07 mm, 0.08 mm, 0.06 mm, 0.05 mm, 0.07 mm, 0.08 mm

All these results have been recorded on a range control chart of the type previously shown in Fig. 3.25. The completed chart is shown in Fig. 3.26. All the parts lie within the control lines so no action needs to be taken.

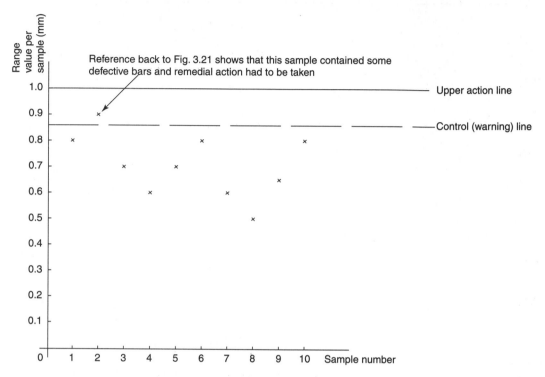

Figure 3.26 Range value control chart for 10 batches of 5000 bars each: inspection sample of 80 bars per batch.

Test your knowledge 3.9

1. Table 3.11 shows the inspection results for a batch of components. Draw a histogram showing the distribution of these results.

Table 3.11

Component size (mm)	Number of components
16 (Lower limit)	0
17	3
18	8
19	16
20 (Basic size)	20
21	14
22	17
23	2
24 (Lower limit)	0

Total sample 80 components

2. Calculate the average value for the data given in Table 3.11.

3. Calculate the range value for the data given in Table 3.11.

3.3.7 Monitoring improvement

All manufacturing companies should be constantly monitoring their products and production methods in order to make improvements. Such improvements should be aimed at cost reduction through the more efficient use of materials and a reduction in the amount of energy required for production. At the same time the aim should also be directed towards improved quality (fitness for purpose).

By making more efficient use of materials and reducing energy consumption, environmental conditions are improved. Environmental improvement should also be a key design aim. For instance, designers of motor car engines should be constantly trying to reduce harmful exhaust gas emissions.

3.3.8 Calibration and checking of control and monitoring equipment

It has already been stated that components cannot be made to an exact size and any dimension must be given a tolerance to allow for an acceptable variation. Similarly, neither can equipment used for measuring, testing and monitoring be made to an exact size. It is generally accepted that test equipment should always be at least 10 times more accurate that the object being measured.

Inspection, measuring and monitoring equipment should itself be regularly tested by comparing it with national and international standards and secondary standards. This can be done in-house if the company is large enough to have its own standards department. Usually, however, the measuring equipment is sent away to specialist laboratories such as the National Physical Laboratory (NPL) for testing and calibration through their United Kingdom Accreditation Service (UKAS). A calibration chart showing the deviation from true size for various settings of the equipment is issued. The calibrated deviations are added to or subtracted from the indicated size, thus converting the indicated size to the true measured size.

Test your knowledge 3.10

1. Explain briefly, in your own words, why it is necessary for all manufacturing companies constantly to monitor:

(a) the performance of their products;
(b) the performance of their plant.

2. An item of measuring equipment is returned from UKAS with a calibration chart.

(a) Explain briefly, in your own words, the purpose of this chart.
(b) If a reading on the instrument is 25.445 mm and the calibration chart shows a corresponding value of −0.006, calculate the actual dimension.

Key Notes 3.3

- Quality assurance is the creation and maintenance of a quality management system to ensure that finished products achieve fitness for purpose and conform to a customer's requirements.

- Quality control is the practical means of achieving quality assurance.
- 'Right first time' is a policy for manufacturing quality products so that wasteful scrap and the need for reworking is eliminated at source.
- Production control aims to achieve the required rate of production at a competitive price so that affordable goods are delivered on time, and that those goods meet the required quality specification.
- Quality standards may be laid down by the customer or by a national organization such as the British Standards Institute (BSI) or an international organization such as the International Standards Organization (ISO).
- Quality specification adopted in the UK is ISO 9000.
- Quality Indicators are the *variables* or *attributes* that control the quality of the finished product and which are capable of being monitored.
- Critical control points. These are those points in the production system where the quality indicators are checked for conformity.
- Sampling techniques may be continuous sampling, random sampling, single sampling or multiple sampling.
- Continuous sampling is only used for key components where lives are at stake. However, for most manufactured goods it would be too time consuming and expensive.
- Random sampling, single sampling and multiple sampling all involve the inspection of a small batch of the finished goods. The overall quality of the goods can then be determined by applying statistical techniques to the results of the inspection.
- Acceptable Quality Level (AQL) is the level of defective components in a batch that the customer is prepared to accept.
- Lot Tolerant Percentage Defective (LPTD) is the point where the percentage of defective components in a batch becomes unacceptable.
- Average control charts plot the results of successive batch inspections so that any trends can indicate whether the process is moving out of control. This enables the machine setter to make adjustments before defective components are manufactured.
- Range control charts are also used for the quality control of production. Both average control charts and range control charts are used to achieve right first time production.
- Monitoring of products and processes is essential to maintain quality products at an affordable price in a competitive manufacturing environment.

Evidence indicator activity 3.3

Write a report in general terms, identifying the key factors of a quality assurance system within a manufacturing organization, supported by:

- a description of the function of quality indicators and critical control points;
- the identification and description of the quality control techniques used;
- an explanation of the role of testing and comparison;
- a description of the subsequent uses of test and comparison data.

The report should be supplemented by an investigation of how quality control techniques are applied at key stages to the production of a product.

Chapter 4

Manufacturing products

Summary   This chapter covers the requirements of Unit 4 of the GNVQ Manufacturing (intermediate). The *unit* is subdivided into four *elements*. The first element (4.1) is concerned with the composition, characteristics and preparation of materials, together with the equipment and machinery used in manufacturing. The second element (4.2) is concerned with the processing of materials and components. The third element (4.3) is concerned with the assembly and finishing of products to a given specification. The fourth and final element (4.4) is concerned with further work on the application of quality control assurance procedures to manufactured goods. This chapter also considers safety and safe working practices, the use of safety equipment, and health and safety procedures and systems.

4.1 The composition, characteristics and preparation of materials

4.1.1 Key characteristics of materials

All materials possess properties or characteristics that make them suitable for use in manufacturing products. These properties or *key characteristics* vary from material to material and include:

- mechanical properties (e.g. strength, toughness and hardness);
- physical properties (e.g. electrical, magnetic and thermal properties together with density, colour and flavour);
- resistance to degradation and/or corrosion.

Generally, the key characteristics of a material are dependent upon its *composition*. Before a material can be selected for the manufacture of a product the following points must be considered:

- the conditions to which the product will be subjected in service;
- the process or processes to which the material will be subjected during manufacture;
- the cost of the material and its processing;
- the service life of the product for which the material is to be used.

Let's now consider the key properties in more detail.

4.1.2 Mechanical properties

The mechanical properties of materials (strength, toughness and hardness) were introduced in Section 3.3.3. We will now consider these and additional properties in more detail.

Strength

In general, we consider *strength* to be the ability of a material to resist an applied force without failure or rupture. In the case of strength, this applied force will be either tensile (stretching) or compressive (squeezing). Alloy steels, composite materials and some non-ferrous alloys are examples of materials that have high strength. As mentioned earlier in this book do not confuse strength and toughness – for example, a hard steel file has a *high tensile strength*, yet it is brittle and can be easily snapped by hitting it with a hammer while it is held in a vice. The file *lacks toughness*. However, if the file is heated until it is red hot and then allowed to cool down slowly (annealed), the file can be hammered and bent without breaking. The file has become *tough* but, at the same time, its *tensile strength* will have been *reduced* by this treatment.

Rigidity (stiffness)

Rigidity or stiffness is the ability of a material to *resist distortion* such as bending and twisting. The frames of machine tools, vehicle chassis and the foundations of buildings require this property. Steel is much stronger than cast iron but it is less rigid. For this reason cast iron is the preferred material for machine tool frames.

Elasticity

Elasticity is the ability of a material to be stretched or deformed and return to its original size and shape. Natural and synthetic rubbers, and metals used for springs and steel rules, are examples of materials and products possessing this property. For elastic materials, the elongation (stretch) is proportional to the load provided they are not overloaded. Double the load and you double the stretch. Halve the load and you halve the stretch. Remove the load and the elastic material returns to its original length.

Plasticity

Plasticity is the opposite property to elasticity. It is the ability of a material to be *permanently deformed* or shaped by the application of a force. Most metals have this property if sufficient force is applied. The steel sheets pressed into the shape of car body panels must have the property of plasticity so that they can retain the required shape when taken from the press. If they were elastic, they would spring back into a flat sheet. Plastic materials (polymers) have this property when heated. This enables them to be moulded to shape.

However, you have to be careful. Metals tend to show elastic properties up to a limiting load (their elastic limit). Beyond this load they either break if they are brittle (e.g. cast iron or quench hardened steel), or they undergo plastic deformation. Thus most metals can show the dual characteristics of elastic or plastic properties depending upon how severely they are stressed. Metals that can show these dual properties include low carbon and medium carbon steel, copper, brass alloys and aluminium and its alloys.

Ductility

Ductility is a special case of plasticity. It is the ability of a material to be *permanently deformed* by stretching. Some metals are very ductile such as the steel used for the manufacture of car body sections by pressing. The metal copper, as used for wire, has to be ductile. This is because the wire is produced by drawing it through dies in order to make it longer and thinner.

Malleability

Malleability is also a special case of plasticity. This time it is the ability of a material to be permanently deformed by a compressive force. Products such as forged crankshafts, camshafts, spanners and small valve bodies are produced from malleable materials. The metals used for the production of bars, sheets and metal ingots by hot and cold rolling also need to be malleable. Most metals become more malleable when they are heated. This is why a blacksmith makes a steel bar red hot before forging it to shape by hammering it on an anvil.

Hardness

Hardness is the ability of a material to resist indentation and abrasion by another hard body. The degree of hardness of a material ranges from that of soft polymers to diamonds. A diamond is the hardest known material. Figure 4.1 shows the relative hardness of a number of common substances.

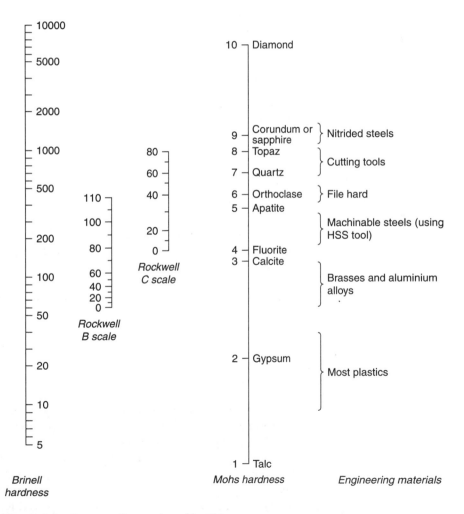

Figure 4.1 Comparative scales of hardness.

Toughness

As already stated, toughness is the ability of a material to *resist impact* or *shock loads*. Hammer heads, cold chisels, springs and high tensile bolts are examples of products possessing the property of toughness.

Brittleness

Brittleness is the opposite of toughness. Brittle materials *lack impact strength*, that is, they *lack of toughness*. Brittleness is often associated with hardness and hard materials. Sintered carbides, quench hardened (untempered) steel, ceramics and glass are examples of brittle materials. Glass breaks easily when hit with a hard object such as a hammer, therefore glass is brittle, it lacks toughness.

Test your knowledge 4.1

1. Name the main mechanical properties required by the materials used to make the following products:

   (a) a steel rule;
   (b) sewing thread;
   (c) a machine frame;
   (d) a squash racket handle;
   (e) an axe head.

2. State the properties that you associate with the following materials:

   (a) concrete;
   (b) glass;
   (c) rubber;
   (d) low carbon (mild) steel;
   (e) nylon.

4.1.3 Physical properties

Electrical conductivity

Electrical conductivity is the ability of a material to conduct electricity. Materials that can conduct electricity are called *conductors*. Most electrically conductive materials are metals, the exception being the non-metal carbon. There is also a group of non-metals called semiconductors used in the manufacture of diodes, transistors and integrated circuits but these are outside the scope of this book. The most common metal used for electrical conductors is copper. It is a very good conductor of electricity, it is easily drawn out into the finest of wires and can be easily soldered to make permanent connections. Other highly conductive metals are aluminium, silver, gold and platinum.

Materials that are *very poor conductors* of electricity are called *insulators*. There is no such thing as a perfect insulating material, only materials that are very poor conductors. Typical materials used for insulators are polymers, composites and ceramics. Dry wood is also an insulator. It is possible, however, to make polymers and elastomers conductive for specialized applications by adding a metallic or carbon powder during compounding. Aircraft tyres are made conductive in this manner so that static electrical charges collected during flying can leak safely to earth on landing. This reduces the risk of sparks that could cause a fire and it also reduces the risk of electric shocks for the passengers as they disembark.

Magnetic properties

Only *ferromagnetic* materials show strong magnetic properties. Such materials are iron, cobalt, nickel and gadolinium. Originally iron and steel were the only magnetic materials available. Since the chemical name for iron is the Latin word *ferrum*, such metals were called *ferromagnetic*. Other metals and non-metals are added to the ferromagnetic materials to modify their magnetic properties as required.

Soft magnetic materials such as pure iron and low carbon steels only become magnetized when placed in a magnetic field. These materials are used for such things as transformer cores and electromagnet cores. They do not retain their magnetism when the energizing magnetic field is removed.

Hard magnetic materials such as quench hardened high carbon steels are more difficult to magnetize when placed in a magnetic field. However, once magnetized, they retain their magnetism even when the energizing magnetic field is removed. Hard magnetic materials are used for the permanent magnets found in loudspeakers. Alloy steels have been developed (such as the alloy Columax) that are as much as 30 times more powerful than hard high carbon steel.

Thermal properties

Thermal conductivity

Thermal conductivity is the ability of a material to conduct heat. Generally the materials most able to conduct heat are the same as those that are most able to conduct electricity. Therefore metals such as copper and aluminium are the most thermally conductive. Such metals are used for the manufacture of cookware and soldering iron bits.

Thermal insulation

Materials that are poor conductors of heat are called *thermal insulators*. Generally, they are the same materials as electrical insulators. Polymers, composites and ceramics are poor or medium conductors of heat. Air is a good heat insulator if it is trapped and cannot circulate. It is the air trapped been two pieces of glass that provides the insulation in double glazing. Again, it is the air trapped in the fibreglass used for loft insulation that prevents heat loss.

Refractoriness

This is the ability of a material to remain unaffected by high temperatures, that is, they can be raised to high temperatures without burning, softening, or melting – for example, the fire bricks that are used to line furnaces are made from refractory materials. Thermal insulators are not necessarily refractory materials. Expanded polystyrene plastic is a very good heat insulator but it melts at quite low temperatures and it will also burn.

Thermal stability

Thermal stability is the ability of a material to retain its shape and properties over a wide or specified temperature range. The metal nickel and its alloys such as 'Invar' are used in the manufacture of measuring tapes and master clock pendulums. 'Monel' is a nickel alloy that can resist high temperatures without appreciable corrosion or dimensional change (creep) and is used for the manufacture of steam turbine blades, and 'K-Monel' alloy is used in the manufacture of pressure gauges.

Density

Density is the mass of material per unit volume of a material. Expressed mathematically:

$$\text{Density}(\text{kg m}^{-3}) = (\text{Mass}) \div (\text{Volume})$$

Manufactured products such as aircraft may be made from materials with a low density such as the aluminium alloys. Other products may also be made from low density materials such as polyethylene. At the other end of the scale, car batteries contain components that are made from lead which has a very high density. Table 4.1 lists the densities of some common materials.

Table 4.1 Density

Material	Density(km m^{-3})
Aluminium	2 700
Titanium	4 500
Lead	11 300
Tungsten	183 920
Polyethylene	960
Polyamide 66	1 150
Polyesters	1 360
Polyvinylchloride	1 400
(PTFE) Polytetrafluoroethylene	2 140

Flow rate

Flow rate is the rate at which a material will flow under pressure and is an important factor in the processing of polymers, health products such as toothpaste, confectionery and food products such as dough, sauces, fondants and pastes. The flow rate of a material will depend upon its viscosity. Viscosity is the ease with which a material will flow – for example, an oil with a high viscosity will not flow easily, while an oil with a low viscosity will flow easily. Viscosity is affected not only by temperature and pressure but also by the consistency and composition of a material.

In the polymer processing industry flow rate is of critical importance in extrusion and moulding processes. Flow rate must also be considered in the pharmaceutical and toiletries manufacturing industries. Products such as toothpaste and ointments would be difficult to squeeze out of the tube if too thick but, if too thin, it would be difficult to control them. This lack of control would result in waste.

Flavour

Flavour is the property of a food or drink determined by taste. The flavouring is the substance, or substances, that gives these products the unique qualities that attract consumers to them. Flavours may be natural such as vanilla and fruit juices or essences distilled from natural products. Nowadays, many essences and flavourings are synthesized, artificial products.

Colour

Colour is the visual effect on the eye produced by light of different wavelengths reflected from the surface of the product. The colouration of the surface of the product will determine the wavelength or mixture of wavelengths reflected. The colour of manufactured products is very important to satisfy customer preference. This is the case particularly in the automobile, textile and clothing industries. Colouring matter is often added to processed foods in order to enhance and improve their colour and to give them a more attractive and appetizing appearance for the consumer. Colouring may be natural such as cochineal, a scarlet colour made from dried insects, or synthetic such as tartrazine made from petrochemicals.

Texture

Texture is the property of a product that is responsible for its 'feel'. Feel is very personal and what may be pleasant to one person might be unpleasant to another. This is particularly the case with fabrics and is important in the clothing and upholstery industries.

Test your knowledge 4.2

1. Before the days of multigrade motor oils, cars were often difficult to start in very cold weather. Explain the reason for this.

2. Calculate the volume of an aluminium casting if its mass is 54 kg. You will find the density of aluminium in Table 4.1.

3. An electric cable consists of copper wire surrounded by PVC plastic. Suggest why these materials were used in its manufacture.

4. In your own words, explain the difference between materials that are thermal insulators and materials that have refractory properties.

5. If a screwdriver blade is stroked with one pole of a magnet, the screwdriver becomes magnetized, and remains magnetized. State whether the screwdriver blade has 'hard' or 'soft' magnetic properties. Give the reason for your choice.

4.1.4 Composition of materials

Composites

Composites are materials composed of two or more different materials bonded together, one being the matrix surrounding the particles or fibres of the other. The particles or the fibres provide the reinforcement. Composites are designed to give properties different from, and superior to, those of the individual materials. Composites can be natural or synthetic. Nature provides some examples of composites that are extremely tough and durable, for example:

- Bone has a hard outer surface with a tough fibrous underlayer and a foam like core (bone marrow).
- Teeth have a hard outer surface called enamel with a tough fibrous inner core called dentine.

- Wood consists of fibrous tubes made from natural cellulose bound together by a natural polymer matrix called lignin as shown in Fig. 4.2.

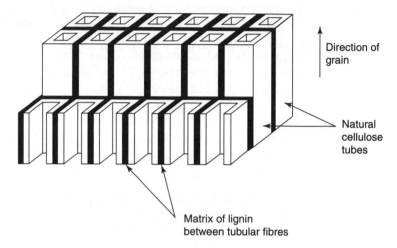

Figure 4.2   Structure of wood.

Synthetic composites also consist of reinforcing fibres or particles bound together by a matrix – for example, glass fibre reinforced plastics (GRP) are used for making a whole range of moulded products from crash helmets to the hulls of boat and ships. The glass fibres of the reinforcing material are bonded together by a synthetic resin such as polyester. Carbon fibre reinforced plastics are used for lightweight high strength mouldings for such products as tennis and squash racket frames, racing car body shells and helicopter rotor blades.

The most common particle reinforced material is concrete where stone chippings are the reinforcing material. The stone chippings are held together in a matrix of a hydraulic cement mortar. More sophisticated particle reinforced materials are used for cermet inserts for cutting tools.

Mixtures

A *mixture* is a close combination of different substances in which no chemical reaction takes place, no new substance is created and no heat is taken in or given out. Further, separation of the ingredients can be achieved using only physical processes – for example, if you mix iron filings and saw dust, you can remove the iron filings with a magnet.

On the other hand, if the reactive metal sodium burns in the poisonous gas chlorine, a new substance called sodium chloride is formed. This substance is the salt used in cooking and as a table condiment. Since a new substance is formed and since heat is given out during the reaction, sodium chloride (table salt) *is a chemical compound*. It can only be split up into sodium and chlorine again by further chemical reactions. Therefore sodium chloride *is not a mixture*.

Concrete is a mixture of aggregate (sharp sand and stone chippings or gravel) and a hydraulic cement. However, once water is added a chemical reaction takes place forming chemical compounds with the cement. It is these reactions that cause the concrete to 'set'.

Mixtures are also to be found in combinations of materials or ingredients held together by physical rather than chemical means such as mixing flour, salt, water and yeast into dough for bread. However, chemical changes do take place during the baking process.

Blending

Blending, on the other hand, consists of mixing different grades of the same substance – for example, the tea in your tea bag will have be a blend of the dried leaves of the tea plants from several different plantations from different locations. They will be selected so that the blend of different flavours will give the required overall flavour at a competitive price. Similarly the coffee you buy will be a blend ground from the roasted beans from a number of different sources. Tobaccos, wines and spirits are also blended in order to improve the flavour or reduce the cost or both.

Alloys

Alloy is a metallurgical term. An alloy is a material composed of a close and complex association of two or more pure metals, or it can be composed of pure metals and non-metals. Such an alloy has its own properties. These are metallic in nature and are superior to, and different from, the properties of its constituents. Therefore alloys are formed in order to obtain a material whose properties are tailored to meet a specified requirement in terms of strength, hardness, toughness, thermal resistance or corrosion resistance. Table 4.2 gives details of some common alloys.

Test your knowledge 4.3

1. State the essential differences between compounds, mixtures and alloys, and give an example of each.

2. Explain briefly what is meant by the term a 'composite material'.

3. Explain briefly the essential difference between an alloy and a pure metal.

4.1.5 Resistance to degradation

All materials will degrade over a period of time when exposed to atmospheric conditions and the rate of degradation will vary from material to material. It is virtually impossible to prevent degradation but it can be retarded by changing the composition of the material (e.g. alloying steel with nickel and chromium to make stainless steel) or by protecting any exposed surfaces (e.g. painting or electroplating). In some cases the degree of degradation can be reduced to a level where it is undetectable by the unaided eye.

Corrosion

Corrosion is the process by which a material is progressively damaged by chemical or electrochemical attack. Most metals are less resistant to degradation by corrosion than ceramics and polymers, and the rates of

Table 4.2 Typical alloys

Alloy	Composition	Properties	Applications
Steel	Iron and carbon	Inexpensive, strong, ductile	Nuts, bolts, sheet strip, plate, rod, general engineering products
Brass	Copper and zinc	Ductile, corrosion resistant	Taps, valves, sheet, bar, rod, nuts, bolts
Bronze	Copper and tin plus a deoxidizer such as phosphorous or zinc	Strong, ductile Corrosion resistant	Marine fittings, bearings, gears
Solder	Lead, Tin, antimony	Low melting point Corrosion resistant	Joining metals Electrical connections
Duralumin	Aluminium, copper, magnesium, manganese	Light, strong, ductile (after solution treatment), Corrosion resistant	Aircraft parts, automobile engine blocks and cylinder heads
Stainless steel	Iron, carbon, chromium, nickel	The properties vary according to composition and heat treatment. All stainless steels are corrosion resistant	Chemical and food processing plant and equipment. Cutlery.
Tool steel	Iron, carbon, tungsten, cobalt, chromium, vanadium	Very hard. Heat and corrosion resistant	Cutting tools for metal, plastic and wood

corrosion vary between metals. Iron and steel are the most common ferrous metals and will corrode (rust) rapidly under the influence of oxygen and moisture when left unprotected. Metals such as brass, bronze and stainless steel are alloyed in such a way as to make them very corrosion resistant, but even so in certain environments (e.g. marine environments) brass will corrode more rapidly than bronze and bronze more rapidly than stainless steel.

Weathering

Weathering is a degradation process in which the material is damaged by exposure to atmospheric effects such as sunlight, rain, heat or cold. Sometimes the effect is physical – for example, if water is absorbed into brick or stone, it will expand as it freezes and split the brick or stone into fragments. Wind and rain will also erode (wear away) the surface of softer materials. Sometimes the sunlight (ultraviolet (UV) rays) will break down the chemical structure of the material. Certain polymers may become embrittled, and they will craze and crack when exposed to sunlight, heat or immersion in water over a long period of time.

Pigments are added to some polymers to act as stabilizers to absorb the (UV) rays in sunlight – for example, window frames are made from a UV resistant grade of polyvinyl chloride (UPVC). In the case of polyethylene (polythene) sheet, carbon black is added to retard UV degradation. In another case, polyethylene detergent bottles are stabilized with polyisobutylene to protect the polymer against the effects of the detergent.

On the other hand, some polymers are deliberately made to degrade. At one time, polymers buried in land fill sites did not break down. Nowadays many polymers are biodegradable (see also Section 1.4.2). They will break down as the result of bacterial attack when buried and not pollute the ground.

Test your knowledge 4.4

1. State the difference between corrosion and erosion.

2. Explain, with examples, where the natural degradation of materials is :

   (a) desirable;
   (b) undesirable.

3. (a) Name three materials that are resistant to atmospheric corrosion.
   (b) Name three materials that degrade rapidly if left unprotected.

4.1.6 Processing methods

There is a wide variety of processes that can be used to manipulate materials and products during manufacture. The processing method selected will depend upon:

- the materials used;
- the size, shape and mass of the product;
- the number to be produced;
- the cost and time constraints imposed by the customer, the market and competitors.

Figure 4.3 shows a flow chart indicating the route that is taken in selecting the method of processing.

4.1.7 Processing methods (heat treatments)

Annealing

Annealing is a process carried out on metals and glass in order to soften the materials, render them less brittle, and remove the effects of previous working thereby increasing their ductility and plasticity. The process consists of heating the material to a given temperature for a set period of time followed by slow cooling, usually in the furnace.

Normalizing

Normalizing is a process mainly carried out on metals to eliminate the stress effects caused by casting, hot and cold working, and welding. The process consists of heating the material to a given temperature for a set period of time followed by slow cooling in still air. This slightly more rapid cooling prevents the excessive grain growth of annealing. The metal will be less ductile but will machine to a better finish. Normalizing is often carried out after rough machining castings and forgings to remove any residual stresses before finish machining. This prevents the components warping after finish machining.

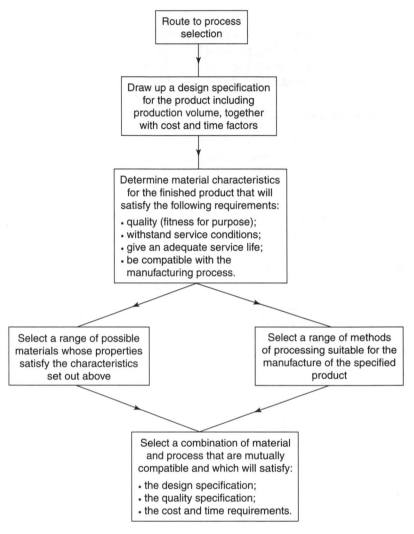

Figure 4.3 Route to process selection.

Stress relief annealing

Stress relieving annealing can only be carried out on metals that have been previously cold worked. It not only removes any residual stresses from previous processing, it also restores the grain formation so that further cold working processes can be carried out on the metal without it cracking. Compared with full annealing, as previously described, it is a low temperature process. Stress relieving is also known as process annealing, interstage annealing and subcritical annealing.

Solution treatment

Some aluminium alloys such as 'duralumin', containing copper, harden naturally under normal atmospheric conditions and have to be softened immediately before use. This softening process is called *solution treatment* since it causes the copper and aluminium to form a solid solution. Solution treatment consists of heating the aluminium alloy to 500°C and allowing it to soak at this temperature for a short while. This renders the alloy soft and ductile and it will still be in this condition when it cools to room temperature.

Hardening

Hardening is a process carried out on metals to improve strength, wear and indentation resistance. The principal methods of hardening are:

- *Cold working* by rolling, drawing or pressing during which the grain structure is distorted and this imparts a degree of hardness. This is the only way of hardening many non-ferrous metals and alloys.
- *Precipitation hardening*. This is a natural process that occurs in certain aluminium alloys, such as 'duralumin', which contain copper. We have just seen that this type of aluminium alloy can be softened by a process called solution treatment. However, in this condition the alloy is unstable and at room temperature the aluminium and copper will slowly combine together to form a *hard intermetallic* compound of copper and aluminium particles that will precipitate out of solution. This is called *precipitation age hardening* and the metal will lose much of its ductility and become harder and more rigid. This process of natural ageing can be speeded up by again heating the metal. Alternatively it can be delayed by refrigeration.
- *Quench hardening* in which medium carbon, high carbon and alloy steels are heated to a high predetermined temperature (red heat) followed by rapid cooling (quenching) in brine, water or oil to give a high degree of hardness. The degree of hardness depends upon the composition of the metal and the rapidity of the cooling. Once the hardening temperature has been reached, there is no benefit in increasing the temperature further.
- *Tempering* is the name of a process carried out after quench hardening to remove the brittleness of the metal and increase its toughness. There will be some small loss of hardness. The metal is reheated to the required temperature and again quenched. Typical tempering temperatures are listed in Table 4.3. The temperature can be judged by the colour of the metal surface after polishing it with emery cloth. Other materials such as glass and some polymers also benefit from suitable tempering processes to reduce their brittleness.

Steam pressing

Steam pressing is a process carried out in the clothing industries for pressing clothing. Steaming is also used in the millinery industry for forming or blocking hats.

Pasteurization

Pasteurization is a process discovered by Louis Pasteur whereby certain food and drink products (particularly milk and milk products) are heated to a high enough temperature to kill any pathogenic organisms and hence make it safe for human consumption.

Table 4.3 Tempering temperatures

Colour*	Equivalent temperature (°C)	Application
Very light straw	220	Scrapers; lathe tools for brass
Light straw	225	Turning tools; steel engraving tools
Pale straw	230	Hammer faces; light lathe tools
Straw	235	Razors; paper cutters; steel plane blades
Dark straw	240	Milling cutters; drills; wood engraving tools
Dark yellow	245	Boring cutters; reamers; steel cutting chisels
Very dark yellow	250	Taps; screw cutting dies; rock drills
Yellow-brown	255	Chasers; penknives; hardwood cutting tools
Yellowish brown	260	Punches and dies; shear blades; snaps
Reddish brown	265	Wood boring tools; stone cutting tools
Brown-purple	270	Twist drills
Light purple	275	Axes; hot setts; surgical instruments
Full purple	280	Cold chisels and setts
Dark purple	285	Cold chisels for cast iron
Very dark purple	290	Cold chisels for iron; needles
Full blue	295	Circular and bandsaws for metals; screwdrivers
Dark blue	300	Spiral springs; wood saws

*Appearance of the oxide film that forms on a polished surface of the material as it is heated.

Ultra-heat treatment (UHT)

Ultra-heat treatment is the process of producing UHT milk by using higher temperatures than pasteurization thereby killing all pathogenic organisms. The process gives the milk a longer shelf life but, unfortunately, it alters the flavour and some people find this unpalatable.

Canning

Canning is a process used in the food processing industry which relies on high temperatures to destroy enzymes and micro-organisms after the food has been placed in the can. Finally the food is sealed in the can to prevent oxidation and recontamination. Canning can also be considered as a packaging process.

Test your knowledge 4.5

1. Choose suitable heat treatment processes for the following purposes:

   (a) softening a quench hardened steel;
   (b) softening a cold rolled brass sheet;
   (c) stress relieving an iron casting after it has been rough machined;
   (d) softening aluminium alloy rivets before use – also state how they can be kept soft;
   (e) hardening and tempering a high carbon steel cutting tool.

2. List the main advantages and disadvantages of the following heat treatment processes as applied to milk: pasteurization, sterilization, ultra-heat treatment (UHT).

4.1.8 Processing methods (shaping)

Moulding and forming

Moulding is a shaping process commonly carried out on polymers and to a lesser degree on glass and clays. Note that polymer materials do not have a defined melting temperature or temperature range like metals. They become increasingly soft and 'plastic' until their viscosity is sufficiently low that they will flow into moulds under pressure. The following techniques being the most widely used.

Compression moulding

Compression moulding is the process of simultaneously compressing and heating a polymer powder, loose or as a compressed heated preform, in a mould. This process is one of the major methods of shaping *thermosetting* plastics. These are polymer materials that undergo a chemical change during moulding. This change is called *curing* and the plastic material can never again be softened by heating. Mouldings made from thermosetting plastics (thermosets) tend to be more rigid and brittle than those made from *thermoplastics*. Compression moulding is suitable for high volume production and may involve the use of positive, semi-positive (flash) or transfer moulds. Sections through typical moulds are shown in Fig. 4.4.

Injection moulding

Injection moulding consists of the injection, under pressure, of a charge of *thermoplastic* material that has been heated to soften it sufficiently to flow, under pressure, into a water cooled mould. Unlike the thermosetting plastics used in compression moulding, thermoplastics soften every time they are heated. Care has to be taken not to overheat them or the plastic will become degraded. The plastic material takes the shape of the mould impression when solidified. Injection moulding is used mainly for the manufacture of thermoplastic products in large quantities. The principle of the process is shown in Fig. 4.5.

Extrusion moulding

Extrusion moulding consists of the continuous ejection, under pressure, of heated thermoplastic material through a die orifice to produce long lengths of formed plastic strip – for example, curtain rail strip. The principle of the process is shown in Fig. 4.6.

Blow moulding

Blow moulding consists of blowing compressed air into a tube of heat softened thermoplastic material clamped between the two halves of a mould whereupon the plastic is forced outwards into the mould impression. The blow moulding process is used to manufacture products such as plastic soft drink and beer bottles, chemical containers and barrel shapes in large quantities. The principle of this process is shown in Fig. 4.7.

Vacuum forming

In vacuum forming, a preheated sheet of thermoplastic material is laid across a mould. The air is pumped out of the mould and atmospheric pressure forces the softened sheet into the shape of the mould. Products such as plastic washing-up bowls are made in this manner. The principle of this process is shown in Fig. 4.8.

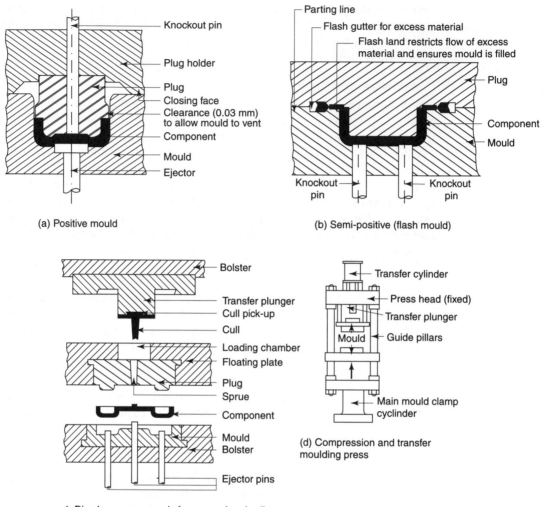

(a) Positive mould

(b) Semi-positive (flash mould)

(c) Transfer mould

1. Die shown open ready for removal and cull.
2. Floating plate closes on moulding powder is loaded into chamber.
3. Transfer plunger descends and forces plasticized moulding powder through spruce into mould.

(d) Compression and transfer moulding press

Figure 4.4 Compression moulding.

Casting

Casting consists of shaping components (castings) by pouring or injecting molten metal into a mould of the required shape and then allowing it to solidify by cooling. Various casting techniques are used depending upon the type of metal, the size and shape of the finished casting and the number of castings required.

Sand casting consists of making a mould by ramming a special moulding sand around a wooden pattern. The pattern is made the same shape as the

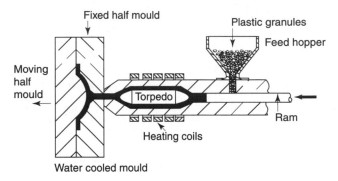

Fixed half mould

Plastic granules

Feed hopper

Moving half mould

Torpedo

Ram

Heating coils

Water cooled mould

(a) Ram injection moulding machine

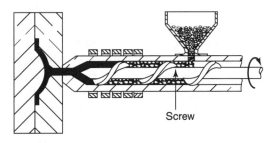

Screw

(b) Screw plasticizer injection moulding machine

Figure 4.5   Injection moulding.

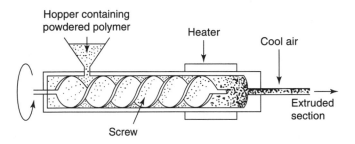

Hopper containing powdered polymer

Heater

Cool air

Screw

Extruded section

Figure 4.6   Extrusion (plastics). Reprinted by permission of W. Bolton.

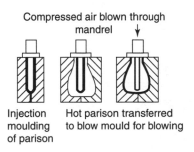

Compressed air blown through mandrel

Injection moulding of parison

Hot parison transferred to blow mould for blowing

Figure 4.7   Blow moulding.

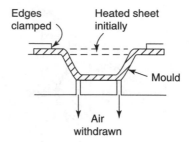

Figure 4.8 Vacuum forming. Reprinted by permission of W. Bolton.

finished product but larger so as to allow for shrinkage of the metal as it cools. The pattern is removed and the molten metal is poured into the mould until it is full. When the metal has cooled and solidified, the sand is broken away to expose the casting. The mould can only be used once. Sand casting is used to manufacture such products and components as bench vice bodies, machine tool beds, motor vehicle engine blocks and cylinder heads. A section through a simple sand mould is shown in Fig. 4.9.

Gravity die-casting uses metal dies in place of the sand mould of the previous process. The dies have to be in two parts so that the casting can be removed when it has cooled. These dies are expensive to manufacture but can be used over and over again. The number of components that can be made before the dies have to be replaced ranges from hundreds to thousands depending

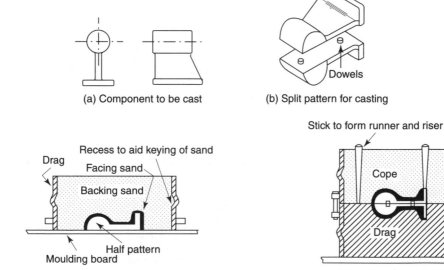

(a) Component to be cast

(b) Split pattern for casting

(c) Lower half of mould (drag) complete

(d) Drag is turned over, top half of pattern is added and the cope is made on top of the drag

Figure 4.9 Sand casting.

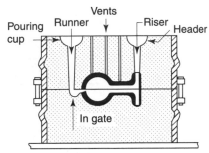

(e) The completed mould is opened and the pattern is removed.
The sticks forming the runners and risers are also removed.
The in gate is cut in the drag and vents and pouring cups are
cut in the cope. The cope and drag are then reassembled
ready for pouring the molten metal

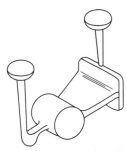

(f) The casting as removed from the mould ready
for the runner and riser to be cut off and any blemishes
trimmed off (fettling)

Figure 4.9 *(Continued)*.

upon the melting temperature of the metal being cast and the alloy steel used
for the dies. This process is used for making castings from metals and alloys
having a low melting point.

Pressure die-casting consists of injecting a charge of molten metal into metal
dies under very high pressure. Again split metal dies are used. The castings
are usually made from metals with a low melting temperature such as
aluminium and its alloys and zinc based alloys such as 'Mazak'. This process
is widely used for the manufacture of car door handles, fuel pump bodies,
badges and other small components. The principle of the process is shown
in Fig. 4.10.

Extrusion

Like plastics, metals can also be extruded. However, the extrusion of metals
involves the use of very much higher temperatures and very much greater
forces. The extrusion process consists of forcing a heated billet of metal
through a die of the desired profile. The extruded section can be in the
form of products of constant cross-section such as tube, rod and complex
profiles such as yacht masts and double glazing window frame sections. The
principle of the process is shown in Fig. 4.11.

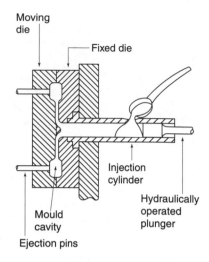

(a) Cold chamber pressure die-casting

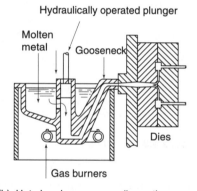

(b) Hot chamber pressure die casting

Figure 4.10 Pressure die-casting.

Forging

Forging is the shaping of very hot metal, by force, in a die. The materials used are carbon and alloy steels. The process consists of heating the metal in the form of a billet to a specified temperature such that it is soft and plastic (but not molten). The heated billet is placed in the bottom half of the die whereupon the top half of the die is brought down at high speed and pressure forcing the metal to fill the die cavity. Forging is used to produce very tough, shock resistant products and components such as crankshafts, camshafts and connecting rods for the automobile industry, and hand tools such as spanners and wrenches. Some details of the process are shown in Fig. 4.12 and the forging temperature ranges for various materials are shown in Fig. 4.13.

Sheet metal pressing

Pressing is the shaping of sheet metal, ferrous and non-ferrous, in a die of the desired shape, under force while cold. The material must be in the annealed

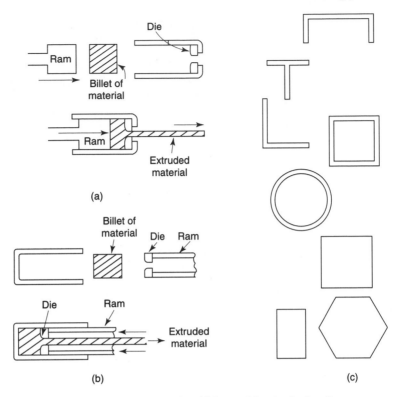

(a) Direct extrusion. (b) Indirect extrusion. (c) Some of the standard sections available with extrusion. Considerably more complex sections are possible.

Figure 4.11 Hot extrusion. Reprinted by permission of W. Bolton.

condition prior to pressing. When a number of pressing operations have to be carried out to achieve the final desired shape the material will invariably require interstage annealing (see Section 4.1.7) to ensure tear- and ripple-free products. Pressing is used to manufacture products such as automobile body parts and domestic appliance casings, oven and kitchenware. A typical press and a pressing are shown in Fig. 4.14.

Test your knowledge 4.6

1. Name a suitable process for manufacturing the following products giving the reason for your choice:

   (a) plastic mouldings for a scale model aeroplane kit;
   (b) a stainless steel sink unit;
   (c) a car engine connecting rod;
   (d) a car engine cylinder block;
   (e) a plastic bucket;
   (f) a fizzy drink bottle;
   (g) plastic hosepipe;
   (h) a plastic case for an electricity meter;
   (i) aluminium alloy pistons for a motor cycle engine;
   (j) lengths of aluminium section for a greenhouse.

2. Describe the essential difference between a thermosetting plastic material and a thermoplastic material.

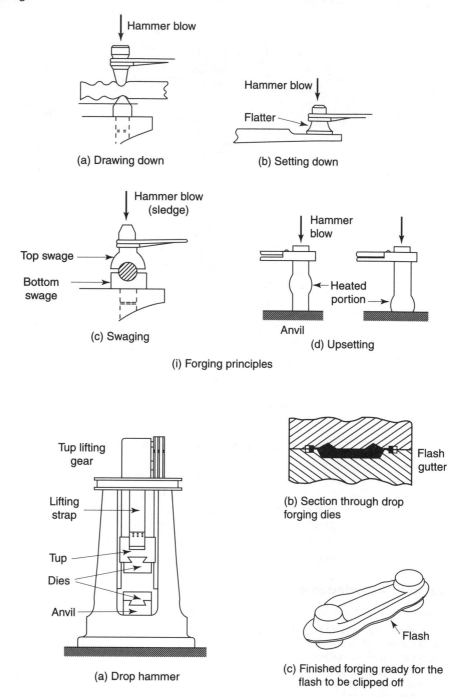

(a) Drawing down

(b) Setting down

Hammer blow

Flatter

(c) Swaging

Top swage

Bottom swage

Hammer blow (sledge)

Hammer blow

Heated portion

Anvil

(d) Upsetting

(i) Forging principles

Tup lifting gear

Lifting strap

Tup

Dies

Anvil

(a) Drop hammer

Flash gutter

(b) Section through drop forging dies

Flash

(c) Finished forging ready for the flash to be clipped off

(ii) Drop forging process

Figure 4.12   Forging.

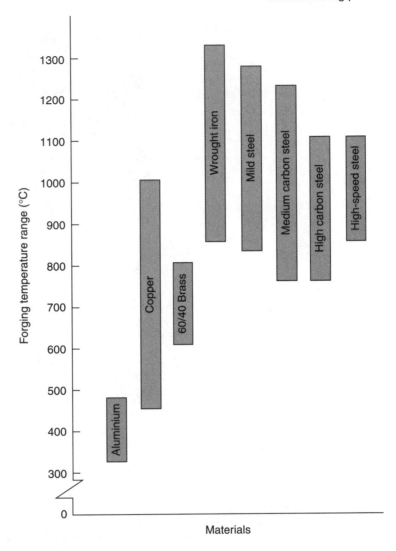

Figure 4.13 Forging temperatures.

4.1.9 Processing methods (surface finishing)

Hot dip galvanizing

In this process, steel components and assemblies are cleaned physically and chemically and immersed in molten zinc to give them a protective coating. A small amount of aluminium is usually added to the melt in order to give the finish an attractive 'bright' appearance. Although the zinc coating is much more resistant to corrosion than steel, nevertheless it will very slowly corrode away in the presence of acid rain and in marine environments. Since, in protecting the steel, the zinc slowly corrodes away it is said to be *sacrificial*. To retard this corrosion even further, galvanized goods may also be painted to protect the zinc finish.

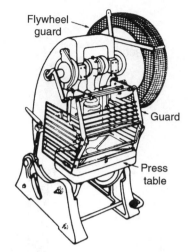

(a) Typical open front power press

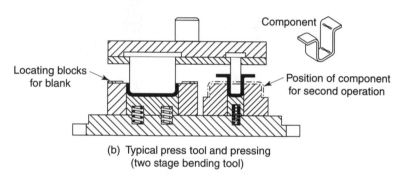

(b) Typical press tool and pressing
(two stage bending tool)

Figure 4.14   Sheet metal pressing.

Conversion coatings

The components to be treated are cleaned physically and chemically and immersed in hot chemical solutions. Complex compounds of metal phosphates, chromates or oxides are formed by conversion on the surface of the metal depending upon the solution used. These surface films not only protect the metal from corrosion directly, they also provide an excellent 'key' for any subsequent painting process. One or other of these processes is frequently used in treating car body panels before painting.

Electroplating

A thin film of a metal is deposited on the surface of the component to be protected by an electrochemical process. The metal film deposited may be intended to prevent corrosion or to impart a more attractive finish or both. The bright parts of a car body trim or a motor cycle are often nickel plated to provide corrosion resistance and then given a further film of chromium to improve the finish. Figure 4.15 shows the principle of the process.

The components to be plated are suspended in a solution called an *electrolyte*. This will depend upon the process being used. The components

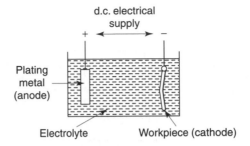

Figure 4.15 Electroplating. Reprinted by permission of W. Bolton.

are connected to the *negative* pole of a direct current electric supply and an *anode* of the metal to be deposited is connected to the *positive* pole. Metal from the electrolyte is deposited on the components to be finished by a process called electrolysis. The anode gradually dissolves into the electrolyte at the same rate as the metal is deposited on the work. Thus the solution strength of the electrolyte is maintained at a constant value.

Anodizing

Although aluminium and its alloys are corrosion resistant to most environments, components made from these metals become dull and grey in time. To prevent this happening or to give them an attractive colour aluminium and aluminium components are often *anodized*. In this process, a transparent film of oxide is deposited on the surface of aluminium or aluminium components. The finish may be self-colour or dyed after treatment. In this process the components to be treated are connected to the positive pole (anode) of an electrolytic cell, hence the name of the finish. The solution of acids will vary according to the finish required.

Plastic coating

Metals may be coated with plastic materials to provide decorative and/or protective finishes. Bathroom towel rails are sometimes finished in this way. Various processes are available but the most widely used are:

- Fluidized bed dipping where the heated components are lowered into plastic powder. The powder is supported on a column of air blown through the porous base of the dipping chamber as shown in Fig. 4.16. This enables a uniform coating to be built up.
- Plastisol dipping, where the work is immersed in a liquid plastisol. No solvents are used and the process provides no toxic hazards to the workers involved.

Glazing

The term glazing means, literally, coating with glass. In fact, the vitreous finish given to ceramic products such as bone china and pottery is exactly this. The ceramic products are given their first firing and this leaves them with a dull porous finish in self-colour. They are then dipped into a slurry of glazing powders suspended in water. It looks rather like a thin mud. The

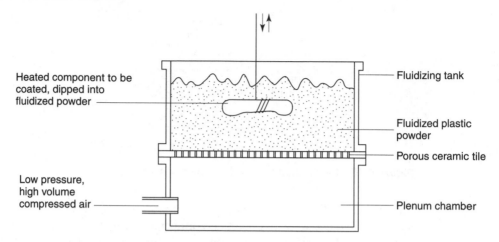

Figure 4.16 Fluidized bed dipping (plastic coating of metal parts).

second firing melts the powders and causes certain chemical reactions to take place. The surface is coated in a glass-like vitreous finish. This *vitreous* finish imparts a colourful, attractive, shiny and impervious finish to these products. Metal products such as roasting tins may also be finished with a *vitreous enamel* to withstand oven temperatures and maintain a smooth, hygienic surface.

The term is also used in many other contexts where a shiny finish is imparted – for example, pastry may be given an attractive brown and shiny finish by applying a beaten egg with a little milk to the surface before cooking. Textiles, papers and cards may also be given a smooth and lustrous finish by treating them with various chemicals.

Laminating

Kitchen unit worktops are often finished by a process called 'laminating'. The wooden, chipboard panels are finished by bonding a thin plastic sheet to them. The plastic sheet not only gives a pleasing colour to the worktop but the plastic is chosen to be heat resistant as well as being smooth and easy to keep clean and hygienic.

Buffing (polishing)

Many metal and leather products are finished by polishing them using a buffing process. Rapidly rotating cloth polishing 'mops' or leather 'basils' are impregnated with a polishing compound and the goods to be finished are pressed against them. Metal components are usually polished in this way before electroplating. They are sometimes given an additional polish after plating as well.

Colouration

Dyes and pigments are either used to colour products directly as in the dyeing of fabrics, or used in surface treatment processes such as painting, lacquering and enamelling. Paints and enamels use pigments and lacquers use dyes.

Dyes are transparent colourants that are soluble. They are derived from natural sources such as insects, berries and wood or from synthetic organic chemicals that are a byproduct of the petrochemical industry. Cloth to be dyed can be treated by immersing it in vats containing solutions of the dyes. The dyed fabrics are often 'fixed' in a second solution to render them colour-fast and prevent fading by exposure to sunlight and washing. Alternatively, the fabrics are woven from yarns that have already been dyed before weaving. Knitwear is usually produced from coloured yarns.

Pigments are opaque colourants and, with exception of carbon black, are derived from natural or synthetic inorganic sources such as metal oxides, metal sulphides or metal carbonates ground into a fine powder. These powdered pigments may be incorporated into plastic moulding materials or incorporated into paints and enamels for surface application.

4.1.10 Processing methods (miscellaneous)

Joining processes

A number of joining processes are used when assembling components. Joining processes will be considered in detail in Element 4.3. They will include such processes as:

- threaded fasteners;
- riveting;
- compression joints;
- adhesive bonding;
- welding;
- brazing, soldering;
- sewing.

Mixing and blending

These have already been introduced as a means of combining or blending various substances together without any chemical reaction taking place. Many food products for human and animal consumption are mixed and blended together, as are some pharmaceutical products. Fertilizers are also mixtures of nitrates, phosphates and potash.

Test your knowledge 4.7

1. Name the finishing process you would use, giving the reason for your choice, for:

   (a) rust proofing a low carbon sheet steel feeding trough for farm use;
   (b) finishing the exhaust system of a motor cycle;
   (c) manufacturing a balanced fertilizer;
   (d) colouring a woven cloth;
   (e) providing a china dinner service with a shiny, impervious and colourful finish.

2. Explain why the colourant used in a paint will be a pigment and not a dye.

3. Explain the essential difference between electroplating and anodizing.

4.1.11 Handling and storage of materials and finished products

In general the location of storage accommodation and areas should be:

- *situated close to the work area* to avoid time wasting and labour intensive handling and transportation;
- *easily accessible* so that the stores personnel do not have to continually use steps and ladders or move one lot of goods to get to another;
- *secure* against unauthorized entry and theft;
- *dry* to avoid damage due to corrosion, staining or decomposition;
- *well ventilated* to provide a pleasant working environment and to avoid any possible build-up of flammable gases;
- *well lit* for easy product identification and avoidance of accidents.

It is very important that materials, components and products are correctly stored and handled to ensure that they are in good condition for processing and in perfect condition for delivery. The major factors that affect handling and storage are:

- radioactivity;
- toxicity;
- oxidation;
- flammability;
- perishability;
- contamination;
- hygiene;
- sharpness;
- discolouration.

Storage areas should not be overcrowded. There should be ample room for handling the products they contain and the storage facilities should be accessible without having to climb about. Appropriate *safety notices* should be displayed where they can be *clearly seen*. Mechanical handling aids should be provided for lifting and stacking heavy objects and the stores staff should be properly trained in the use of such equipment. Stored goods must not obstruct emergency exits.

Radioactivity

Radioactive isotopes are used for taking X-ray photographs of welded joints on pressure vessels and high pressure pipe runs. These isotopes emit highly dangerous gamma rays. They must only be used and handled by suitably qualified persons and must be stored in accordance with the appropriate guidelines and legislation. They must be securely stored against theft and misuse and kept separate from all other materials.

Toxicity

Toxicity is the property of a material or substance to have a poisonous effect on the body by touching, swallowing or inhaling. Toxic materials may be liquid, solid, powder or gaseous in form, and should be stored in a cool, dry and well-ventilated and drained place. Toxic materials can affect our bodies in two ways:

- *Irritants* affect our flesh both externally and internally. Externally irritants can cause skin complaints such as sensitization and industrial dermatitis. More serious is the internal irritation caused by poisons that can affect the major organs of the body. Such irritants can cause inflammation, ulceration, poisoning and the formation of cancerous tumours.
- *Systemics* affect the fundamental organs and nervous systems of our bodies. They affect the brain, the liver, the kidneys, the lungs and bone marrow. Systemics can lead to chronic and disabling illness and early death.

Narcotics, even in small concentrations, cause drowsiness, disorientation, giddiness and headaches. Such effects seriously affect the judgement of a worker and his or her ability to carry out assigned tasks correctly or to control machines and equipment. In larger concentrations narcotics can produce unconsciousness and death.

Toxic and narcotic substances include degreasing chemicals, solvents for paints and adhesives, liquid petroleum gas process chemicals used in surface treatments and heavy metals such as lead, cadmium and mercury. They also include the pesticides, herbicides and organophosphates used in the agrochemical manufacturing industries. The storage procedures and regulations laid down for each substance must be rigidly adhered to; these may include storage on pallets, temperature control and limits on the number of containers of a material that can be stacked on top of each other.

Toxic and narcotic liquids should be stored off the ground to aid speedy leak detection and improve the ventilation and dispersion of any dangerous vapours. The appropriate regulations relating to the stored substances should be strictly enforced and inspections carried out regularly, and each inspection and its findings logged.

All toxic and narcotic substances should be handled with extreme caution and suitable protective clothing worn as necessary. Containers and packaging should be secure at all times and clearly indicate their contents. They must be handled with care and be kept in an undamaged condition so that no spillages or leaks occur.

Oxidation

Oxidation is a chemical reaction with the gas oxygen. Oxygen gas is an important constituent of the air we breathe. Metals, plastics and foodstuffs are all subject to deterioration by oxidation on exposure to the atmosphere. On some metals, especially in the presence of heat and/or moisture, oxygen will cause corrosion, that is, the surface of the metal becomes oxidized. Plastics may degrade due to the effects of oxygen and sunlight, especially outdoors, unless they are stabilized in some way. The environmental stress cracking (ESC) of some polymers is liable to occur when in the presence of moisture or immersed in liquids.

Food will spoil due to increasing bacteriological action in the presence of oxygen at room temperature over a period of time. It is therefore very important that in all the above cases storage must be by such means as to reduce the effects of oxygen to a minimum.

Flammability

Flammability is the ability to combust or burn. Burning is a very rapid oxidation accompanied by the generation of heat. Flammable materials should be stored in cool, dry, well-ventilated and well-drained accommodation in which all statutory regulations concerning such materials are rigidly enforced. Flammable materials include the following:

- gases;
- plastics;
- lubricants.
- paints;
- timber;
- fuels;
- fabrics;
- solvents;
- paper;
- chemicals;
- explosives;

All these substances must be stored and handled with great care, ensuring that there is no risk of sparks or naked flames in their vicinity. It is mandatory that they are all kept in non-smoking areas away from direct sunlight. Many of these substances are subject to local authority regulations and also come within the jurisdiction of the local fire prevention officer. Special insurance requirements may also be enforced.

Perishability

Perishability is the property of a material to perish, spoil or rot over a period of time and this applies mainly to foodstuffs and rubber goods such as tyres, gloves and aprons. Foodstuffs can be kept fresh and deterioration retarded by using a combination of the correct storage temperature and packaging materials. The packaging materials will protect the product from physical damage and keep within its own micro-environment. The correct storage temperature will maintain freshness over the designed shelf life (the period of time for which the product can be displayed before being purchased). Table 4.4 gives the industry recommended temperatures for the various types of food storage area.

Table 4.4 Storage temperatures

Environment	Temperature range
Dry food store	10°C to 15°C
Cold stores	3°C to 4°C
Refrigerators	−1°C to 4°C
Freezers	−20°C to −18°C

Contamination

Contamination is the inadvertent mixing due to contact of one substance with another. Raw foodstuffs are particularly prone to contamination from each other. This contamination may simply be strong smelling and strong flavoured foodstuffs affecting the taste and smell of more delicately flavoured foodstuffs by contact or close storage – for example, meat, fish, cheese, cut garlic and onions. More seriously is the risk of infection when foodstuffs such as raw meats are in contact with cooked meats and other foods not requiring cooking.

Liquids, powders and granules are also prone to contamination and must be stored and handled with great care prior to packaging. Detailed attention must be paid to their method of containment for sale. Compression moulding powders are often pressed into a preform or pellet prior to moulding to avoid the risk of contamination, to aid handling and to ensure correct filling of the mould.

Hygiene

Hygiene is the maintenance of cleanliness of persons and equipment and is of prime importance when handling and storing food, medicines and pharmaceuticals. In all instances personal hygiene is particularly important. Food processing machinery must be capable of being cleaned and serviced relatively easily. Suitable lighting and ventilation is also a factor in maintaining hygiene.

Food processing operatives and handlers must keep every part of their person that comes into contact with food, clean. In the case of hands cuts and sores must be covered with a waterproof dressing or non-permeable gloves. A more responsible approach is to redeploy a person with a cut, sore or infection until it has healed. Clean and regularly laundered industrial clothing suitable for the task should also be worn.

Any infectious or contagious illness contracted by persons involved in the preparation and production of processed foods will necessitate their being taken off production until a doctor has certified that they have the necessary clearance to return.

Sharpness

Sharp or rough edged and abrasive materials should be handled with caution and suitable protective clothing or equipment must be worn when handling them. Materials in this category include:

- paper;
- laminates;
- bone.
- cardboard;
- glass;
- timber;
- metals;
- aggregates;
- gravel;
- stone;
- plastics;

A selection of suitable gloves and 'palms' for hand protection are considered in Section 4.1.15.

Discolouration

Discolouration is a change in colour or pigmentation due to the effects of natural light, in particular that of the ultraviolet (UV) and infrared (IR) wavelengths. It can also be caused by long-term immersion in liquids such as detergents, water, brine solutions and seawater. Reaction with atmospheric oxygen can also cause discolouration. Materials subject to fading or discolouration such as plastics products, stained or painted surfaces, fabrics and timber products should not be stored outside where they would be subject to weathering and direct sunlight. Such conditions cause loss of colour due to changes in the pigmentation (fading).

Products such as wall coverings, carpets and furniture (indoor products) will also fade and discolour when exposed to sunlight. This can be avoided by covering any windows with a thin film of tinted plastic sheet to reduce the intensity of light and filter out the harmful rays. Most commercial and industrial stock is stored in artificially lit storerooms. Some products such as earthenware pipes for rainwater (stormwater) and sewage are stored out of doors, in the open and are not subject to fading or discolouration because these products have become light stabilized during manufacture. They are, however, susceptible to frost damage.

Shelf life

A product, or a component, can be affected by one or more of the factors previously discussed. This makes it necessary for a range of precautions to be taken. In addition to controlled storage conditions, the length of time in storage is also important. Many products, and particularly foodstuffs, may slowly deteriorate however carefully they are handled and stored. Therefore their storage time must be limited to ensure they reach the customer in prime condition. This time is called the product's *shelf life*. The 'sell by date' should be clearly marked on the product and/or its packaging. This does not necessarily mean the products have become unsafe. In the case of preserved foodstuffs it may only mean that the colour, texture and flavour are below standard. Biscuits and crisps may start to become soft after their 'sell by date' has passed. Products such as batteries may have a shortened service life if they are old stock.

Test your knowledge 4.8

1. List the essential requirements for the despatch stores of a fertilizer manufacturer. Pay particular attention to easy handling and keeping the products in good condition.

2. Explain briefly the difference between toxic and narcotic substances and name an example of each.

3. Explain briefly how metal goods, wood products and foodstuffs can be protected against degradation by oxidation and damp.

4. Name the three conditions that are essential for burning to take place.

5. Explain, briefly, the difference between personal hygiene and equipment hygiene and how processing hygiene may be achieved.

4.1.12 Preparation of materials and components

An important stage in manufacturing is the preparation of materials and components prior to processing. Preparatory stages can be difficult to identify when it can be said that all stages in manufacture are steps along the road to the finished product. This is especially so in the food industry where at any stage during manufacture the product cannot be said to be an individual component of the whole – for example, the grain from a field of wheat is the finished product for the farmer but only the raw material for the miller; the flour from that grain is the finished product for the miller but only the raw material for the baker. Similarly, in engineering, a subassembly may consist of many individual components such as nuts, bolts, washers, screws and springs, etc. All these are the finished products for their individual manufacturers, but only the starting point for the manufacturer of the subassembly. Again, the subassembly is only one part of the complete assembly.

There are many different ways in which materials and components are prepared. However, in the interests of quality control and the avoidance of production delays, there are a number of checks that should be made no matter what materials are involved and no matter what is to be manufactured from them. All materials need to be checked for:

- type (is it the correct material?);
- quality (is it to specification?);
- damage (has it been damaged or blemished prior to or during delivery?);
- size (is the material the size ordered?);
- quantity (has the correct quantity been ordered so that production can proceed?).

Once the raw materials, bought-in components and subassemblies have passed their acceptance checks, their preparation for production can commence. The preparation of materials and components ready for processing can include such activities as those shown in Fig. 4.17. The preparatory processes in this figure are not manufacturing sequences but merely lists of possible processes.

One way of illustrating the preparation of materials and components is by means of flow charts for various products as shown in Fig. 4.17.

Test your knowledge 4.9

1. Draw up flow charts for two examples of material preparation for products of your own choice. The products should be from different sectors of manufacturing industry.

4.1.13 Preparation of equipment and machinery

Prior to the manufacturing process all equipment, machinery and tools must be correctly and thoroughly prepared. Apart from safety considerations this is very important in the manufacture of quality products. There are two categories of machinery, *manually operated* and *automated*. First let's consider manually operated machines.

Manual (non-automated) machinery

Machines that are operated by a machinist or an operator are said to be *manually controlled* (non-automated). This method of control depends upon the skill and concentration of the operator and can lead to variations in product quality. The sense of sight of the operator is the major part of his control system and in a company employing *total quality management*, the operator inspects, gauges or measures the components or products produced on his or her machine. Manual machines can be linked to form an integrated manual production system. The relationships between the inputs, outcomes and information flow for a manually controlled machine are shown in Fig. 4.18.

Automated machinery

Automated machinery is not new. The mechanization of the manufacture of rigging tackle was introduced early in the nineteenth century when the

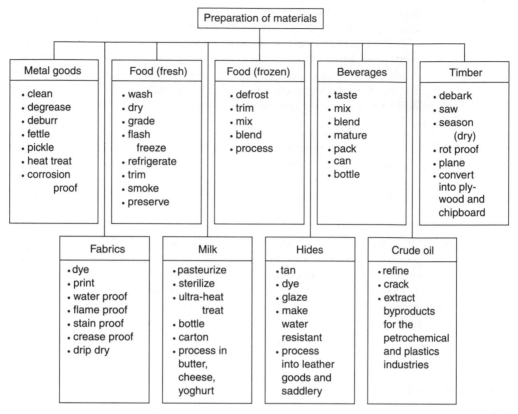

Figure 4.17 Examples of material preparations.

Royal Navy depended on sailing ships. Automatic screw manufacturing machines controlled by steel cams were also introduced in the nineteenth century. Mechanically controlled automated machines were and still are widely used throughout all sectors of manufacturing industry. It is only recently, with the development of powerful microcomputers, that the use of computer numerically controlled (CNC), automated machine tools linked with industrial robots has become so widespread.

Computer numerically controlled (CNC) machines are controlled by a dedicated microcomputer. The operator has to program the computer using the data on the component drawing. The machine is then set and adjustments made for any small variations in tool and cutter sizes. Once this has been done the machine will produce identical parts of identical quality. It will not get tired and if it is fitted with automated equipment to load and unload it (industrial robots) it can carry on working in the dark when everyone has gone home. The relationships between the inputs, outcomes and information flow for automated (CNC) machines are shown in Fig. 4.19.

There are a number of advantages in using computer numerical control:

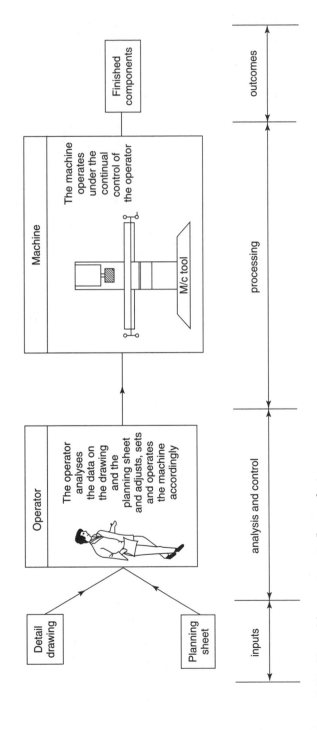

Figure 4.18 Manual (non-automated) manufacture.

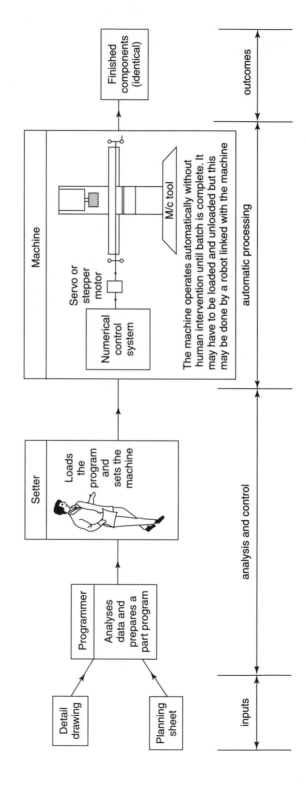

Figure 4.19 Automated manufacture.

- The program can be stored on disk or tape for reuse when repeat batches are required.
- The writing and loading of a program is much quicker than manufacturing control cams of complex design.
- There is total management control over performance, quality and costs.
- The machine control system can be easily linked with other computer-controlled equipment and systems.

With computer controlled equipment it is possible for a number of machines to be interconnected to master computer to form an integrated automated production unit.

This type of machinery requires high capital investment but give the benefits of maintained high output with a consistent level of product quality. Built-in or *in-process* measuring systems constantly monitor the size, shape or other parameter of the product at each critical control point and provide feedback to a comparison unit which provides correction automatically when there is a drift towards the predetermined, specified limits.

Preparation of equipment and machinery

The first task to be considered in the preparation for manufacture is the space around the equipment and machinery. This can be determined before the machinery or equipment is installed by the use of the scale models that are available for most standard machine types and sizes. Whole workshops can be planned, three dimensionally, in this way.

Space

A machine must have adequate space around it so that the operator can move around freely in order to operate it. The movement of features such as slides, tables, arms and heads must also be considered in order to make sure that they do not encroach upon the room needed by the operator at the extreme limits of their travel. Such features must also be checked to make sure that they do not encroach upon the space needed by adjacent machinery. There must be sufficient space for easy delivery and removal of work that may require mechanical lifting and transport facilities.

Floor

The floor space surrounding machinery and equipment should clean, even and uncluttered. Substances such as cutting fluid, oil, grease or fat could cause an operator to slip and sustain an injury if left on the floor.

Lighting

It is essential that operators can clearly see all the machine's controls, screens, dials and gauges and on manual machines the item undergoing processing. As well as good overall lighting the machine or equipment should have its own lighting system that can be positioned by the operator to illuminate the point of processing.

Stability

Machinery and equipment must be positioned in such a way that it will neither move nor vibrate unduly during normal operation. If bench mounted, then the bench must be strong enough to support the weight of the machine and the work. It must be rigid enough to be free from vibration and stable so that it cannot overturn. Normally a machine or piece of equipment is secured to the floor by bolts or similar device. These provide a means of permanently anchoring machinery and equipment to the floor. The machines are levelled by means of wedges. However, some machines that are inherently stable are free standing and supported on adjustable, anti-vibration mounts.

Cleanliness

Ensuring that machinery and equipment is clean and free from dirt is a fundamental part of preparation for processing, and is an essential action in the manufacture of food products in the interests of hygiene. Clothing and upholstery manufacture also requires the utmost cleanliness if the products are to be kept clean, unmarked and fit for sale. Before any cleaning takes place make sure that the machine is switched off at the mains and if necessary post a warning notice on the switchbox. Do not rely on interlock switches or machine controls. Always follow the manufacturers' instructions on cleaning the machinery.

Safety

Ensure that all safety devices are in place and available. In particular, ensure that all guards are in position, secured and in a serviceable condition. Never operate machinery or equipment without the guards in position. If there is any doubt about the condition and/or function of safety equipment and guards, notify your supervisor immediately. Do not operate the machine or equipment until the fault has been corrected. Guards must only be set and/or adjusted by a suitably qualified person and must not be tampered with by the operator.

Tooling

If it is required, make sure that the correct tooling is available for the operation to be carried out and that it is in a serviceable condition. Should the tooling be damaged or blunt it must be returned to the stores for replacement or refurbishment. The use of tools that are incorrectly selected or in poor condition will result in loss of production and poor quality products. Further, the use of such tools can be dangerous.

Settings

All controls and instruments must be set correctly for the process to be carried out. Settings will include speeds, feeds, depth of cut, approach and run-out distances, cycle times, processing temperatures and thread tensions in sewing machines. Let's now have a look at some machine settings.

Worked examples

Output

A confectionery manufacturing company makes chocolates in a range of 10 varieties. They are all processed on the same conveyor belt, 10 abreast. The confection forming the centre of each chocolate is placed on the conveyor belt with their bases predipped in chocolate. The chocolates are placed in line on the belt, are 20 mm long and spaced 20 mm apart, and are to be chocolate coated and decorated in three operations before setting as they pass along the conveyor. The chocolates are to be packed 10 to a box, one of each variety. Calculate how many boxes of chocolates are produced per hour when the conveyor belt runs at 4 metres per minute.

Since the data is given in both millimetres and metres we must choose one or the other. Let's settle on all dimensions in millimetres.

Boxes produced per hour = (conveyor speed × unit time) ÷ chocolate spacing

Where:

$$\text{space taken by one chocolate} = 40 \, \text{mm}$$

$$\text{speed of conveyor belt in mm} = 4000 \, \text{mm/min}$$

$$\text{unit time} = 1 \, \text{hour(60 min)}$$

Therefore:

Boxes of chocolates produced per hour = 4000 mm/min × 60 min ÷ 40 mm

Boxes of chocolates produced per hour = **6000**

Speed

A batch of components is to have a 20 mm diameter hole drilled in each. The recommended cutting speed is 30 metres per minute. Calculate the spindle speed to which the machine must be set.

Use the following formula to calculate the speed:

$$N = (1000 \times S) \div (\pi \times d)$$

Where:

$$N = \text{spindle speed in rev/min}$$

$$d = \text{drill diameter in millimetres}$$

$$S = \text{cutting speed in metres/min}$$

$$\pi = 3.142$$

Substituting the following values in the formula:

$$S = 30 \, \text{m/min}, \, d = 20 \, \text{mm}, \, \pi = 3.142$$

Spindle speed$(N) = (1000 \times 30) \div (3.142 \times 20)$ rev/min

Spindle speed$(N) = $ **477 rev/min** (to the nearest whole number)

Feed

If the feed rate for the previous example is 0.02 mm/rev, calculate the time taken for the drill to penetrate a component 15 mm thick plus 8 mm allowance for the drill point.

$$T = (60P) \div (N \cdot F)$$

Where:

$$T = \text{time in seconds}$$
$$P = \text{depth of penetration}$$
$$N = \text{spindle speed in rev/min}$$
$$F = \text{feed rate in mm/rev}$$

Let:

$$P = 15\,\text{mm} + 8\,\text{mm} = 23\,\text{mm}$$
$$N = 477\,\text{rev/min (from previous example)}$$
$$F = 0.02\,\text{mm/min}$$

Then:

$$T = (60 \times 23) \div (477 \times 0.02)$$
$$= \textbf{147 seconds} \text{ (to the nearest whole number)}$$

Cycle time

A batch of components is to have a hole punched in each using a piercing tool mounted on a press. The cycle time for each component consists of the following elements:

3 seconds handling component
2 seconds placing component
3 seconds punching component
2 seconds removal of component
2 seconds cleaning the piercing tool

Therefore the cycle time for punching the hole in each component is the *sum of the above elements*:

$$3 + 2 + 3 + 2 + 2 = \textbf{12 seconds}$$

Test your knowledge 4.10

1. Briefly compare the advantages and limitations of automated and manually operated machine tools.

2. Summarize the main factors that must be considered when setting up either:

   (a) a circular saw to cut an 8 ft by 4 ft plywood panel into 100 mm wide strips;
   OR
   (b) a sewing machine to make a batch of 20 shirts.

3. Calculate the time taken to drill a hole in a metal component if: the drill diameter $= 14\,\text{mm}$; the cutting speed is $33\,\text{m/min}$; $\pi = 22/7$; depth of penetration $= 12\,\text{mm}$ (ignore point allowance); rate of feed $= 0.04\,\text{mm/rev}$.

4. Calculate the cycle time for turning 1000 components on a CNC lathe given the following data:

Load chuck	5 s
Face end and turn outside diameter	24 s
Chamfer	3 s
Drill	15 s
Bore	30 s
Unload chuck	5 s

4.1.14 Preparation of tools

Tools fall into three general categories:

- cutting;
- shaping;
- assembly and fixing.

Cutting tools

Cutting tools are used to remove surplus material during manufacture. Some trimming tools may also fall into the cutting tool category. The materials from which cutting tools are made must have the properties of hardness in order to maintain a keen cutting edge, but with sufficient toughness to prevent chipping. They must be easily resharpened or disposable. Starting with the simplest cutting tools, Fig. 4.20 shows a selection of shears, scissors and knives used for cutting out and trimming.

Scissors

Scissors are used to cut textiles, plastics, paper, card, metal foil, woven glass fibre, glass fibre mat and food. The correct type being selected for the process to be carried out. They should be checked for sharpness and cleanliness especially those with inserted blade sections. Scissors range in size from the very small as used for needlework, up to the very large size (shears) used in industries incorporating textiles in their products. While scissors would be used for cutting out hand-made garments, for example bespoke tailoring, power driven rotary shears are used for mass produced clothing with several thicknesses of cloth being cut at the same time.

Knives

Knives are used for cutting and for trimming such materials as food, plastics, wood, card, glass fibre roving and leather. The correct type of knife should be selected for the process to be carried out. Knife blades should be kept clean and sharp and when not in use should be covered or sheathed to prevent damage to the cutting edge and for safety. Knives may have blades that are fixed, adjustable, retractable or disposable.

Shears

Shears or 'snips' are used for cutting out thin sheet metal such as tinplate. They are available with straight and curved blades and come in a variety of sizes. For thicker materials bench shears are used as shown in Fig. 4.21(a) and for the thickest sheet and thin plate power driven guillotine shears are used.

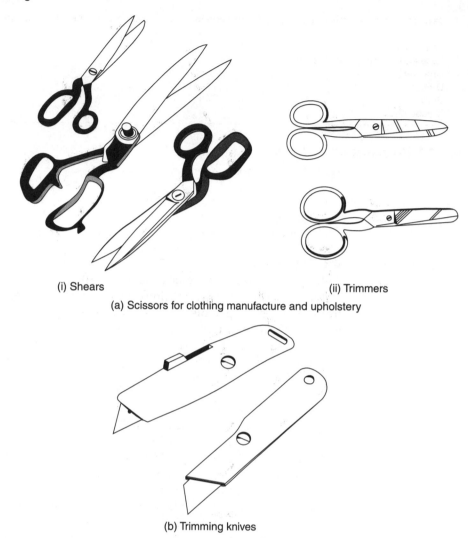

(i) Shears   (ii) Trimmers

(a) Scissors for clothing manufacture and upholstery

(b) Trimming knives

Figure 4.20   Scissors and knives.

Guillotine shears

Guillotine shears are used to cut sheet materials such as metal, plastics, paper, card and composite material and may be treadle operated or power driven. Figure 4.21(b) shows a typical power driven guillotine shear. Before using this or any machine, the operator should be fully conversant with the controls, ensure that all the guards are in position and know the emergency stop procedure. On some power driven machines a photoelectric cell is positioned in front of the blade to prevent operation if any object breaks the beam. For efficient use the blades should be kept sharp and the clearance between the blades should be adjusted to suit the type and thickness of the material being cut.

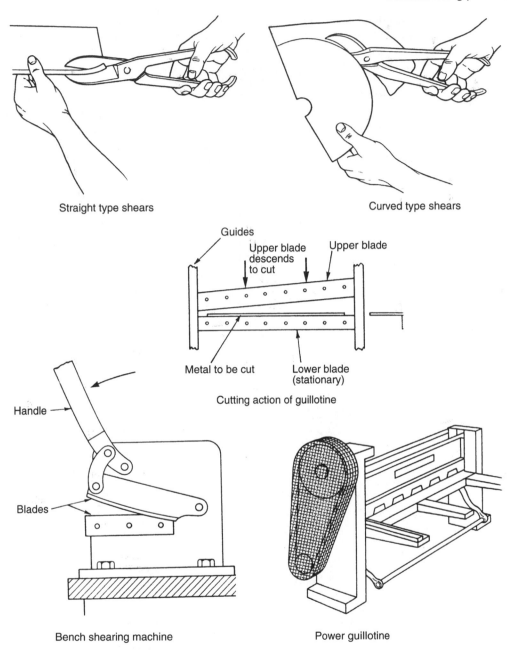

Straight type shears

Curved type shears

Guides

Upper blade descends to cut

Upper blade

Metal to be cut

Lower blade (stationary)

Cutting action of guillotine

Handle

Blades

Bench shearing machine

Power guillotine

Figure 4.21 Sheet metal cutting tools.

Hacksaws

Hacksaws are used to cut metal, plastic, composite materials, frozen meat and bone, and may be hand or power operated. The blades used in hacksaws can be changed to suit the job in hand. It is important to select the correct type and size of blade for the task to be carried out. Hacksaw blades are classified by the pitch of their teeth and length, and are selected according to the thickness of the material to be cut. They range in size from 150 mm long for a 'junior' hacksaw blade up to 600 mm for a power hacksaw blade, the most common size being 300 mm for a hand hacksaw.

Carpenters' saws

Saws are used to cut wood, wood products such as chip and block board and some composite materials. The correct type and size must be selected for the task to be carried out. Saws are classified by size and type, and may be selected according to the pitch of the teeth to suit the thickness or width of the material to be cut. Typical saws are shown in Fig. 4.22(a).

Files

Files are mainly used on metals to shape, deburr and finish surfaces. They can also be used on other materials such as plastics, composites and hardwoods. Files should be selected for the process to be carried out according to their shape, grade, size and length. They are classified by the shape of their cross-section, for example flat, hand, round, half round, square and three

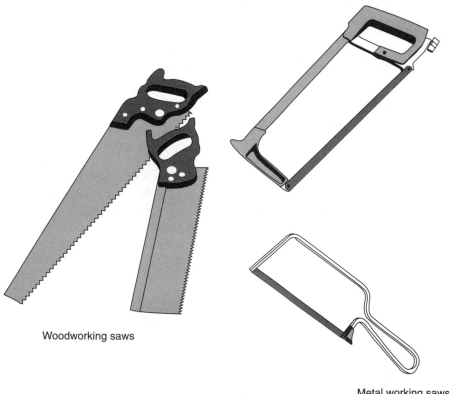

Woodworking saws

Metal working saws

(a) Typical saws

Figure 4.22 Saws and files.

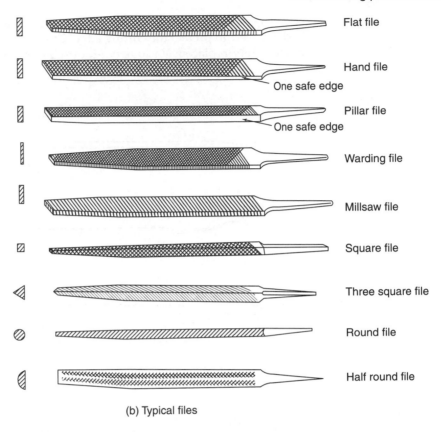

(b) Typical files

Figure 4.22   (*Continued*).

square (triangular) in the following grades of cut: coarse, bastard cut, second cut, smooth and dead smooth. For very fine work needle files and riffler files are available.

Before use files should be checked for serviceability and cleanliness. The handle should be properly fitted, free from cracks, and the ferrule should be in place. A properly fitted handle prevents the pointed tang of the file piercing your hand or wrist, with serious consequences. A cracked or split handle should be immediately replaced. Any metal or dirt clogging the teeth should be removed with a file card (a special type of wire brush). Clogging can be greatly reduced by rubbing chalk into the cutting faces. Also, due to their extreme hardness and consequent brittleness, files should not be tapped on hard surfaces to remove swarf (particles of cut metal) as they may break or fracture. Some typical files are shown in Fig. 4.22(b).

Scrapers

Scrapers as used in cabinet making and the leather goods industry are used to remove rough grain, burrs and sharp edges. Fitters in the engineering industries use scrapers to produce flat surfaces, or make shafts and bearings fit each other. The types of scraper available are flat, half round and three square (triangular) and should be selected to suit the task in hand. Scrapers should be kept clean, sharp and free from grease. When not in use the blade should be cased to protect the cutting edge. Cracked or split handles should be immediately replaced. Various types of scraper and typical applications are shown in Fig. 4.23.

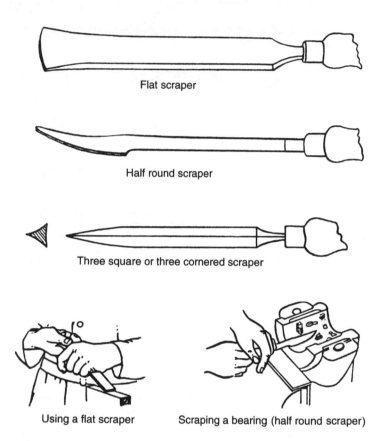

Flat scraper

Half round scraper

Three square or three cornered scraper

Using a flat scraper Scraping a bearing (half round scraper)

Figure 4.23 Scrapers.

Drills

Drills are used to cut holes in materials such as metals, wood, plastics, brick, masonry, glass and composites. They are made of high carbon or high speed steel. Drills used to cut holes in brick, masonry and glass have cutting edges in the form of a tip of sintered carbide to resist the highly abrasive effects of these materials when cut. The correct size and type of drill should be selected for the hole to be drilled, and should be sharp and clean. The drills are rotated by hand or, more usually, by electrically or pneumatically powered drilling machines. These may be fixed or portable. Always ensure that electrically powered drilling machines are checked for electrical safety before use, for example, ensure that the plug is secure, and that the cable is not cut, frayed or twisted. Eye protection in the form or safety spectacles or goggles should always be worn when drilling. Chuck guards are not required on portable drilling machines but must be fitted to floor or bench mounted drilling machines. Some typical drills and associated cutters are shown in Fig. 4.24.

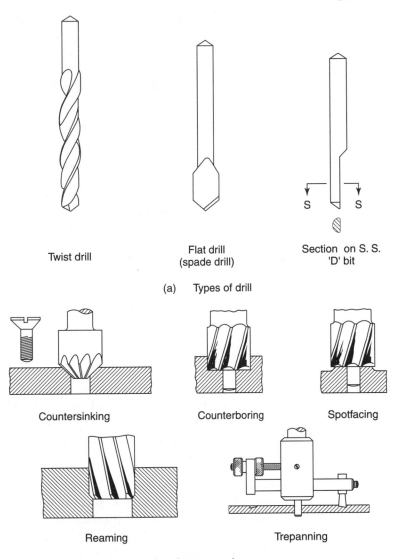

Twist drill

Flat drill
(spade drill)

Section on S. S.
'D' bit

(a)   Types of drill

Countersinking

Counterboring

Spotfacing

Reaming

Trepanning

(b) Associated cutters and processors

Figure 4.24   Drills and associated cutters.

Abrasive paper and cloth

Although not strictly a cutting tool, abrasive papers and cloths have been included in this section since they remove small amounts of material by a cutting action. The cutting action is provided by particles of hard abrasive material such as 'emery' (a form or aluminium oxide) bonded to paper or cloth. Abrasive papers and cloths are used to produce a smooth finish on wood, plastics, metals, glass, stone, painted surfaces and composite materials. The correct type and grade should be selected for the process to be carried out, for example glass paper should be used on wood and plastics, with emery papers and cloths being used on metals.

The types of abrasive papers and cloth available include glass paper that can be supplied in nine grades with abrasive grain sizes ranging from 180 (finest) to 50. Water proof paper sheets (wet and dry) can be supplied in 17 grades with abrasive grain sizes ranging from 1200 (finest) to 60. Emery cloth sheets and strip supplied in 10 grades with abrasive grain sizes from 320 (finest) to 40.

Shaping tools

Some tools used for shaping materials have already been introduced. These are tools used to change the shape of materials without cutting them – for example, forging dies, extrusion dies and plastic moulds. The tablet forming devices used in the pharmaceutical industry can also be considered as a type of shaping tool. Typical blacksmith's tools and processes were shown in Fig. 4.12. Some typical sheet metal forming tools are shown in Fig. 4.25.

Fixing and assembly tools

Fixing and assembly tools are used on components and products to join, assemble or secure in some way during the manufacturing process. Some typical examples are shown in Fig. 4.26.

Screwdrivers

Screwdrivers, as their name implies, are used to drive screws into materials such as wood, plastics and metal. The correct type and size of screwdriver should be chosen to suit the head of the screw being driven – for example, the types of screwdriver available include flat blade, 'Phillips', 'Posidrive', socket and various other shapes introduced by the computer industry. Screwdrivers should be regularly checked for damaged blades and the condition and security of the handle.

Spanners

Spanners are used to turn items such as nuts and bolts when assembling and securing components and products made of wood, metal and plastics. The correct size and type of spanner must be used to prevent bolt heads and nuts being damaged and to prevent possible injury to the user. The types of spanner available include open-ended, combination, ring, box, adjustable and socket. Socket spanners must be used with a wrench that may be of the ratchet type or fixed.

Hammers and mallets

Hammers and mallets are mainly used on metals and wood. The correct type should be selected for the material and process to be carried out. Hammers are classified by the shape of the head or head feature and to a lesser extent weight and include ball pein, cross pein, straight pein, claw, planishing, creasing, stretching and blocking. Mallets are classified by their face material and size and include boxwood, hide, lead, aluminium, zinc, nylon, copper, rubber, polyurethane and polypropylene. Before use hammers and mallets should always be checked to ensure that the head is secure and the handle clean, dry, free from grease and in good condition.

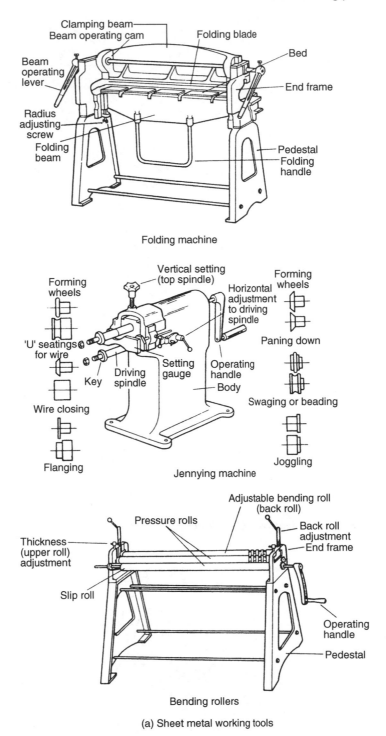

(a) Sheet metal working tools

Figure 4.25 Typical sheet metal forming tools and operations. From *Basic Engineering Craft Course Workshop Theory (Mechanical)* by R. L. Timings, Longman. Reprinted by permission of Pearson Education Ltd.

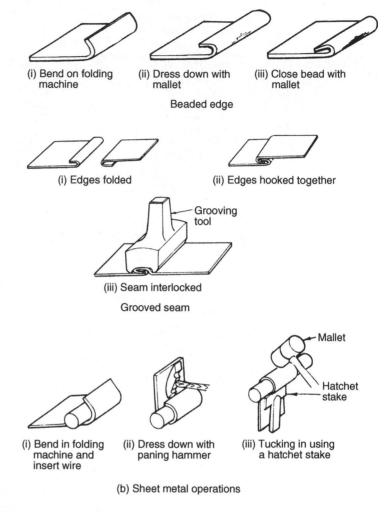

(i) Bend on folding machine   (ii) Dress down with mallet   (iii) Close bead with mallet

Beaded edge

(i) Edges folded   (ii) Edges hooked together

Grooving tool

(iii) Seam interlocked

Grooved seam

Mallet

Hatchet stake

(i) Bend in folding machine and insert wire   (ii) Dress down with paning hammer   (iii) Tucking in using a hatchet stake

(b) Sheet metal operations

Figure 4.25 (*Continued*).

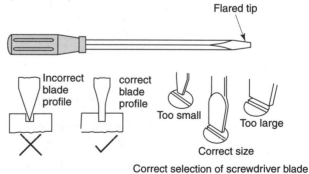

Flared tip

Incorrect blade profile   correct blade profile   Too small   Too large

Correct size

Correct selection of screwdriver blade

(a) Flat blade screwdriver

Figure 4.26 Fixing and assembly tools.

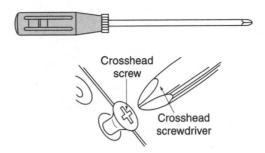

The correct type and size of crosshead screwdriver must always be used for crosshead (recessed-head) screws

(b) 'Posidrive' type screwdriver

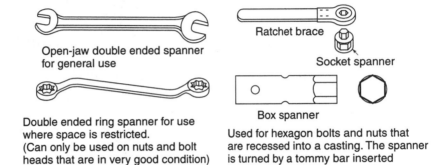

Open-jaw double ended spanner for general use

Ratchet brace

Socket spanner

Double ended ring spanner for use where space is restricted. (Can only be used on nuts and bolt heads that are in very good condition)

Box spanner

Used for hexagon bolts and nuts that are recessed into a casting. The spanner is turned by a tommy bar inserted through holes in the spanner body

(c) Spanners

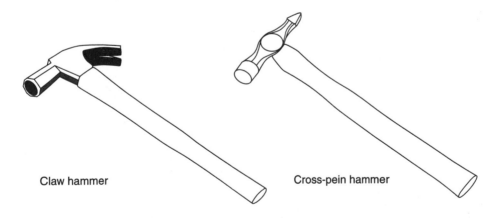

Claw hammer

Cross-pein hammer

(d) Carpenters' hammers

Figure 4.26 (*Continued*).

(e) Engineers' ball-pein hammer

(f) Stove heated soldering irons

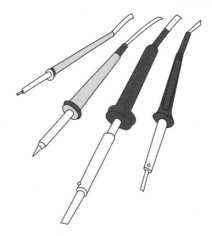

(g) Electrically heated soldering irons

Figure 4.26   (*Continued*).

Soldering irons

Soldering irons may be of the traditional gas heated type which are used mainly on sheet metal and allied products, or electrically heated which are used mainly on electric and electronic components and assemblies. The gas heated type requires a separate heat source (soldering stove). For each type of soldering iron the correct size and or power rating should be chosen for the task to be carried out. Both types should be inspected for serviceability and safety prior to use, with the bit being clean and secure, as should be the handle. The wiring and connections should also be checked.

Riveting tools and guns

Rivets are used to join sheet metal, plastic, composite materials and components by means of a hand or powered riveting gun. Hand riveting can be carried out with a ball-pein hammer and riveting snaps. These are special tools shaped to form the rivet head. Hollow or pop rivets can be fitted by hand using a special tool. This pulls a headed mandrel into the plain side of the rivet to form a head and hence secure it, the protruding part of the mandrel is then broken off by the riveting 'gun' and the joint is made. These riveting techniques are considered in Section 4.3.2. Some POP®* riveting guns have a magazine of rivets to save the operator having to load and place the rivets one at a time.

Riveting guns are usually pneumatically powered and may be used for cold riveting small steel rivets and non-ferrous rivets of copper or aluminium. Hot riveting is mainly used on the large rivets required in structural steel work, bridge building and ship building were it is largely giving way to welding. In this process the rivets are of steel. They are raised to red heat before being placed in the hole ready for forming the head (closing the rivet) in order to secure the rivet in place.

Needles

Needles are used for hand or machine stitching textiles, leather and canvas as used in the clothing, bedding and upholstery industries. The correct type and size of needle should be selected for the task to be carried out. Needles for hand stitching are available in 26 sizes from number 1 which is long and fat to number 26 which is short and thin.

Needles used for industrial sewing machines are classified by the shape of their points and their cross-sections. Point shapes for piercing needles may be sharp, rounded or ballpoint, the body having a circular cross-section. Point shapes for cutting needles are mostly spear point, and the cross-section shape may be lozenge, triangular, or flattened oval, the body having a circular cross-section.

The major difference between domestic hand and machine sewing needles and those used on industrial machines is that the industrial needles have the eye at the point end and the body has a groove running along it to the eye.

---

* POP® is a registered trademark of Tucker Fasteners Ltd.

Test your knowledge 4.11

1. Select suitable tools and equipment for making single prototypes of the following articles:

   (a) a tinplate tray 250 mm × 100 mm × 25 mm deep with soldered corner joints OR the link shown in Fig. 4.27;
   (b) a garment of your own choice;
   (c) a garden bird table OR a nesting box.

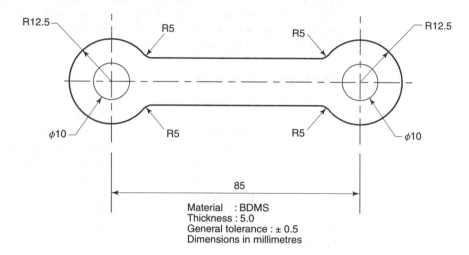

Material : BDMS
Thickness : 5.0
General tolerance : ± 0.5
Dimensions in millimetres

Figure 4.27   Link.

2. Describe how you would check and prepare the tools used in (1) above.

3. Explain briefly why it is bad practice and possibly dangerous:

   (a) to use a spanner that is a poor fit on the nut;
   (b) extend a spanner to obtain more leverage.

4.1.15 Safety equipment, health and safety procedures and systems

It is said that safety is 'everyone's business'. This is particularly true in all fields of manufacturing. It is most important that safety equipment, procedures and systems are inspected and checked regularly to ensure that they are in place and functioning whenever and wherever a manufacturing process takes place.

Emergency equipment

Emergency equipment may include fire extinguishers, hose reels, axes, fire blankets and sprinklers. Examples of these devices are shown in Fig. 4.28 and will be described in greater detail below. It is essential that such equipment is checked regularly and kept in good condition. Such checks must be logged in a register. If an extinguisher is used, it must be recharged without delay as soon as the fire has been put out. A fire certificate may also be required for premises depending on the type of business and the number of people employed.

Fire extinguishers

For combustion to commence and to continue three things are necessary: *heat*, *oxygen* (air) and *fuel*. If any one of these is removed the fire will go out. The

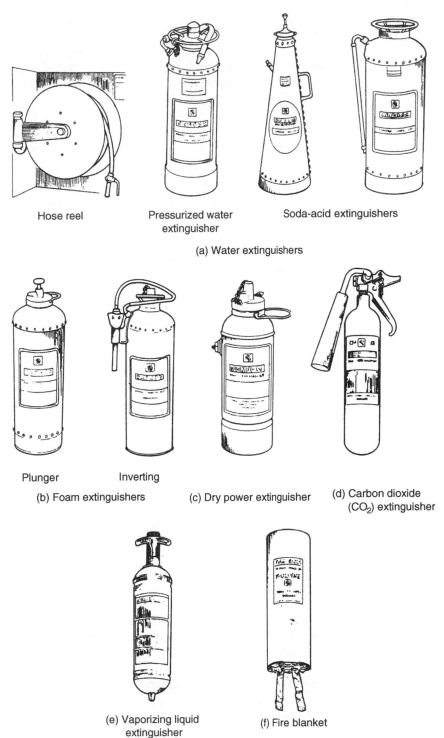

Hose reel

Pressurized water extinguisher

Soda-acid extinguishers

(a) Water extinguishers

Plunger

Inverting

(b) Foam extinguishers

(c) Dry power extinguisher

(d) Carbon dioxide (CO_2) extinguisher

(e) Vaporizing liquid extinguisher

(f) Fire blanket

Figure 4.28 Typical fire-fighting equipment.

purpose of fire extinguishers is to prevent air getting to the burning materials (fuel) and also to cool them down. They may be adequate for putting out small, localized fires or containing fires until the professional brigade can arrive. Remember that human lives are more important than property. If you are using an extinguisher always make sure you have an escape route in case the fire spreads and gets out of control.

Fire extinguishers are colour coded to indicate their contents and the class of fire on which they are to be used as listed in Table 4.5. The classes of fire are as follows:

- Class A. Fires involving *solid materials*, usually of organic nature, in which combustion normally takes place with the formation of glowing embers, for example, wood, paper and textiles, etc.
- Class B. Fires involving *liquids* such as oil, fat, paint, etc., and liquefiable solids.
- Class C. Fires involving *gases*.
- Class D. Fires involving *metals*, magnesium, sodium, titanium, zirconium.

Table 4.5 Fire extinguishers

Type of extinguisher	Colour code	Class of fire
Water	Red	A
Foam	Light Cream	B
Dry chemical powder	Blue	D
Carbon dioxide (CO_2)	Black	Electrical fires
Vaporizing liquid	Green	A,B,C and electrical fires

Pressurized water extinguishers

Water extinguishers are used on wood, paper and textile (class A) fires. The water from the extinguisher cools down the burning material as well as smothering the fire with the steam produced. The extinguisher contains water and a pressurized carbon dioxide (CO_2) canister. When the extinguisher is operated, the CO_2 gas is released and forcibly ejects a stream of water through the nozzle. In a way, it is like a giant soda water siphon.

Foam extinguishers

Foam extinguishers are used on class B fires involving flammable liquids such as oils, fats, paint, petrol and solvents. The foam forms a blanket over the fire thus excluding the air and smothering the fire. The extinguisher contains two chemical solutions that become mixed on operating the extinguisher. The two chemical solutions react together to produce a large volume of foam that is forcibly ejected from the nozzle by the pressure generated. Foam and pressurized water extinguishers must not be used on fires associated with electrical equipment. The foam and the water is electrically conductive and the operator of the extinguisher could receive a fatal electric shock.

Dry chemical powder extinguishers

Dry chemical powder extinguishers are also used on flammable liquid fires, indoors and outside. The non-toxic powder (sodium bicarbonate) is easily removed making it ideal for kitchen and food store use. This type of extinguisher can also be used on class D fires and on fires involving electrical equipment since the powder is non-conductive. The heat of the fire causes a chemical change in the powder and large amounts of carbon dioxide gas are given off. This further blankets the fire.

Carbon dioxide extinguishers

Carbon dioxide (CO_2) extinguishers are used on electrical appliance and flammable liquid fires. The carbon dioxide gas released from the extinguisher excludes the oxygen and smothers the fire. There are two precautions to note when using this type of extinguisher. First, if the fire cannot breathe, neither can you. As you use the extinguisher, back away from the bubble of CO_2 gas, do not advance towards it. Second, this type of extinguisher is of little use out of doors or in a draughty position indoors, as the cloud of CO_2 gas is blown away before it can smother the fire.

Vaporizing liquid extinguishers

Vaporizing liquid extinguishers are used on motor vehicle and electrical equipment fires. These extinguishers have the advantage that a small amount of the liquid used will produce a very large volume of vapour when heated. Vapour concentration must exceed 9 per cent by volume to be effective and is thus best suited to automatic systems as fitted in machine rooms, motor vehicle workshops and computer suites. Like the CO_2 extinguisher, they displace atmospheric oxygen. If the fire cannot breathe, neither can you. The same rules apply as for using a CO_2 extinguisher.

Hose reels

Hose reels are fixed fire-fighting appliances connected directly to a pressurized water supply with a shut-off nozzle fitted at the end of the hose. The nozzle is often adjustable to give either a sustainable and powerful jet of water, or a fine spray, to fight class A fires. A spray is often better than a jet as it does not scatter the burning material and spread the fire. Further, the spray flashes into steam more easily and blankets the fire as well as cooling it. The reach of the hose is limited to the length of hose on the reel, this is usually 25 m to 40 m.

Fire blankets

Fire blankets are useful in all fields of manufacturing and are made of fire resistant synthetic fibres. They are used to smother a fire or wrap around a person on fire to extinguish the flames. In a food preparation area, they can easily be thrown over a burning chip pan to exclude the air. They do not contaminate the food.

Sprinklers

Sprinklers are now installed in many manufacturing and warehousing premises in the UK. Sprinkler systems have sensors that detect smoke or heat from a fire and automatically operate to release a spray of water onto the fire in order to extinguish it. Being automatic they have the advantage of operating when the premises are unattended.

First aid equipment

Minor and major injuries are an ever-present occurrence in the manufacturing industries due to negligence, carelessness and the flouting of safety rules, no matter how stringently the requirements of the HSE are implemented. Therefore we need to consider the first aid equipment that must always be available in a manufacturing plant. The basic equipment consists of:

- first aid kit;
- stretcher;
- blankets;
- protective clothing.

The Health and Safety Regulations (First Aid) 1981 specify the contents of first aid kits and boxes, and that these must be checked at regular intervals and replenished as necessary. Any out-of-date contents must be discarded. A number of people, sufficient for the number of employees in a company, factory site or workshop, must be trained and hold an appropriate first aid qualification approved by the Health and Safety Executive.

The contents of the first aid box are specified according to the number of people in a particular work area, the minimum requirements for 1 to 10 persons are as follows:

- One guidance card.
- Twenty individually wrapped sterile adhesive dressings (assorted sizes) appropriate to the working environment. They must be coloured so that they are visibly detectable in the catering industry.
- Two sterile eye pads, with attachment.
- Six individually wrapped triangular bandages.
- Six safety-pins.
- Six medium sized, individually wrapped, sterile, unmedicated wound dressings (approx. 10 cm × 8 cm).
- Two large, sterile, individually wrapped, unmedicated wound dressings (approx. 13 cm × 9 cm).
- Three extra large, sterile, individually wrapped, unmedicated wound dressings (approx. 28 cm × 17.5 cm)
- Six moist cleaning wipes individually wrapped.

The approved code of practice issued with the regulations stipulates that an employer should provide a suitably equipped and staffed first aid room in premises where 400 or more employees are at work or in cases where there are special hazards. In cases where employees are using potentially dangerous machinery or equipment, portable first aid kits must be provided. Every accident that occurs in the workplace no matter whether it is a minor or a major incident must, by law, be entered in the accident book.

Personal safety clothing and equipment

There is a wide range of personal safety clothing and equipment available. Items such as safety footwear are designed to protect workers from harmful external hazards such as materials or equipment falling on their feet. Other items such as gloves and hats worn by workers in the food industry are used to protect products from contamination. Some examples of personal safety clothing and equipment are shown in Fig. 4.29.

Personal safety clothing and equipment includes the following items:

- Approved coveralls.
- Safety footwear.
- Eye protection.
- Safety helmets.
- Respirators.
- Gloves and gauntlets.
- Ear protection.
- Hats and caps.

Approved coveralls

Approved coveralls are primary protection garments, and should be close fitting, in good condition and have no buttons missing or unfastened that could cause them to become entangled in machinery or other equipment. The purpose of overalls varies according to the industry – for example, in the engineering industry overalls protect the wearer, and his or her clothes, from oil and dirt. However, in the food industry overalls are worn to protect the food from contamination as well as protecting the wearer. Therefore in food preparation areas it is important to wear clean, washable, light coloured

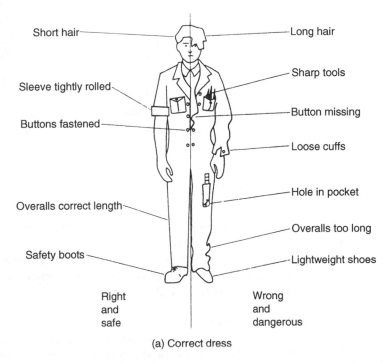

(a) Correct dress

Figure 4.29 Safety clothing and equipment. Courtesy of Training Publications Ltd.

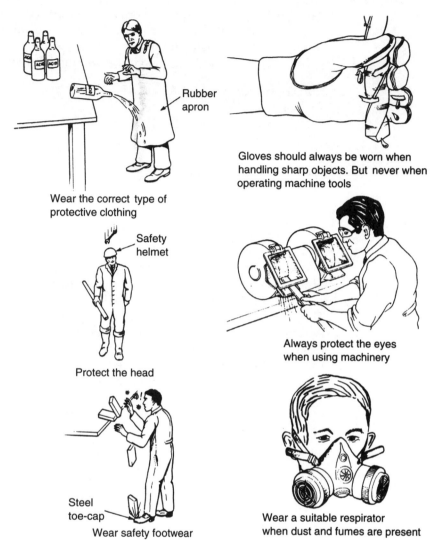

Rubber apron

Wear the correct type of protective clothing

Gloves should always be worn when handling sharp objects. But never when operating machine tools

Safety helmet

Protect the head

Always protect the eyes when using machinery

Steel toe-cap

Wear safety footwear

Wear a suitable respirator when dust and fumes are present

(b) Additional protection

Figure 4.29 *(Continued)*.

protective clothing which completely covers ordinary clothing, or is worn in place of it. Coveralls in the form of a boiler suit, warehouse coat or laboratory coat are worn to protect food from the risk of contamination by such things as fabric fibres and hair. The condition of coveralls should be checked regularly, particularly for tears and loose buttons or fastenings.

Safety footwear

Safety footwear includes boots, shoes, overshoes and slipovers and should include one or more of the following features which are also shown in Fig. 4.30:

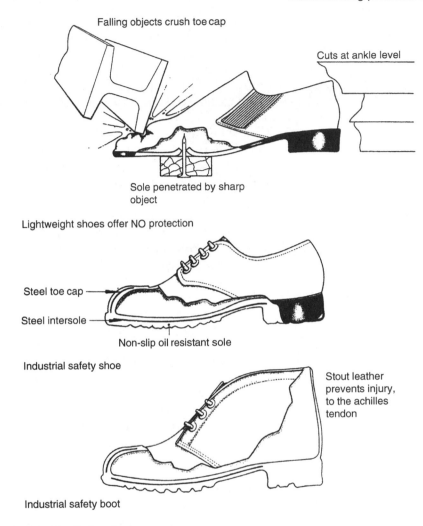

Falling objects crush toe cap

Cuts at ankle level

Sole penetrated by sharp object

Lightweight shoes offer NO protection

Steel toe cap

Steel intersole

Non-slip oil resistant sole

Industrial safety shoe

Stout leather prevents injury, to the achilles tendon

Industrial safety boot

Figure 4.30 Safety footwear.

- steel toe caps;
- oil and solvent resistant soles;
- heat resistant soles;
- anti-static soles;
- sole plates;
- anti-slip soles.

Regular checks should be carried out to ensure that the footwear is in good condition and gives adequate protection, for example the soles should not be worn smooth and the steel toe caps must be free from dents or other damage.

Eye protection
Due to the many different types of eye protection available it is vital to wear the correct type of protection for the work to be carried out.

Safety spectacles with side guards provide adequate protection for general industrial use and, depending on the impact rating of the lens, may also be used when machining or handling small quantities of chemicals. In more hazardous environments, for example where there is a possibility of molten metal or chemical splashes, spark showers or projectiles, safety goggles, full face visors or face shields should be worn. These are made of special safety glass or plastic.

Hand-held or helmet type screens fitted with the appropriate dark glass filter should be worn when arc welding in order to protect the eyes from ultraviolet (UV) light as well as the heat and excessively bright visible light radiated from the weld pool. Goggles fitted with filter screens should be worn when gas welding. Gas welding goggles only give protection from heat and excessively bright visible light, as no ultraviolet rays are present. Therefore gas welding goggles are unsafe and unsuitable for arc welding (see Section 4.3.3). Regular inspection of eye protection should be carried out to ensure that its serviceability is maintained.

Safety helmets

Safety helmets (hard hats) should always be worn when there is a risk from falling objects, or when entering low structures, as when working around scaffolding. The wearing of such headgear is compulsory in these areas and access will be refused if this rule is not complied with. Regular checks should be made to ensure the serviceability of helmets and that they have no dents or cracks and that the lining is intact. The harness of safety helmets must be adjusted to the individual wearer in order to give the correct clearance. For reasons of hygiene hats should not be passed from one person to another, and the harness should be regularly washed and cleaned.

Respirators

Several different types of respirator are available, ranging from the simple dust mask that traps fibres and particles of material, to the canister, cartridge, valve and vapour types which absorb organic, inorganic vapours gas and trap particles. Air fed coverall suits give full protection from vapours, toxic gases and fine particulate material. As with all safety equipment respirators must be checked and tested regularly for serviceability and canisters or elements replaced after the required period.

Gloves

Several types of protective glove are available for different working environments and some examples are shown in Fig. 4.31. The materials from which they are made and the purposes for which they are worn are listed in Table 4.6. Note: Worn and torn gloves should be discarded. They offer little protection and can become entangled with machinery.

Ear defenders

Ear defenders may be of the disposable type made from soft or mouldable material so that they can fit snugly in the ear. Alternatively they may be of the type designed to cover the whole ear and are worn rather like headphones.

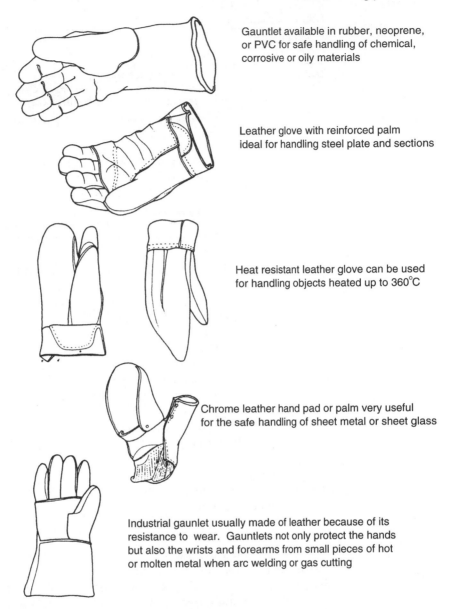

Gauntlet available in rubber, neoprene, or PVC for safe handling of chemical, corrosive or oily materials

Leather glove with reinforced palm ideal for handling steel plate and sections

Heat resistant leather glove can be used for handling objects heated up to 360°C

Chrome leather hand pad or palm very useful for the safe handling of sheet metal or sheet glass

Industrial gaunlet usually made of leather because of its resistance to wear. Gauntlets not only protect the hands but also the wrists and forearms from small pieces of hot or molten metal when arc welding or gas cutting

Figure 4.31 Gloves. From *General Engineering* by R. L. Timings, Longman. Reprinted by permission of Pearson Education Ltd.

Ear defenders are worn as protection against noises that affect concentration and cause impairment to hearing. For reasons of hygiene, they should never be swapped between people and must be checked, cleaned, disinfected and tested regularly.

Hats and caps

Hats and caps may be worn for two purposes. They may be worn for personal protection to prevent hair from becoming entangled in machinery.

Table 4.6 Glove materials

Type of glove material	Use
Leather	Handling hot and rough objects in heat treatment, foundry and welding areas. Asbestos gloves should no longer be used as asbestos is a health hazard
Armoured (reinforced) leather	Handling sheet metal, sharp objects and sheet glass
Rubber	Handling chemicals and particularly liquids such as acids, alkalis, lubricants, cutting fluids and toxic liquids generally
Thin plastic	Handling foodstuffs
Thick plastic	Handling swarf and metal offcuts, particularly if contaminated with dirt, oil and grease

This type of accident can result in serious, scalping type injuries. Hats and caps are worn in the catering and food processing industries to protect the product being manufactured, for example processed foodstuffs, from becoming contaminated with loose hairs and/or dandruff.

Hazardous environments

Appropriate safety clothing must be worn in all hazardous environments, so let's consider what constitutes a hazardous environment. Hazardous environments are those which may singly or in combination be:

- unpleasantly or dangerously *hot*;
- unpleasantly or dangerously *cold*;
- contaminated;
- physically dangerous;
- electrically dangerous;
- hygienically dangerous.

Hot environments

Hot environments include foundries, heat treatment, moulding and metal dip coating shops, bakeries and confectionery processing plants. Any form of clothing will give some protection from heat. However, in the hot environments just described, thick clothing made of insulating materials will delay the penetration of heat to the body. Such clothing must be made from flame proof or flame retardant materials. Most synthetic fibres are unsuitable since they melt when heated and may stick to the skin causing burns that are difficult to treat.

Metallized suits and aprons give greater protection against radiant heat. Gloves will protect the hands against heat and hot materials, but the gloves must be made from a material to suit the material being handled or process being carried out. Prechilled garments, such as ice vests, are worn as protection against excessive heat. Air or water cooled suits can also be used but they require external air and water services in the form of trailing hoses that limit mobility and radius of operation in intensely hot environments. Under extreme conditions a proprietary brand of protective clothing known as a 'Vortex' suit can be worn to protect the body.

Cold environments

Cold environments include refrigeration plants and cold stores for meat and food products. Solution treated aluminium alloy components (see Section 4.1.7) also have to be kept under refrigerated conditions to delay precipitation age hardening. Cold environment clothing, necessary to prevent hypothermia and frostbite, must not only provide thermal insulation but also allow the evaporation of perspiration thus avoiding it freezing on the body; these requirements cannot be provided by a single material. Several layers of protection are preferable so that individual layers of clothing can be removed to suit the environment and the bodily heat generated while working.

Contaminated environments

Working in contaminated environments, such as those associated with paint spraying and where toxic, asbestos and radioactive materials are present, require the use of high levels of protection. Full protective (safety) clothing for such hostile environments includes respirators, gloves and masks. Respirators are used to filter out harmful particles, gases and fumes but only in oxygen bearing atmospheres. Breathing apparatus with its own air supply such as that worn by firemen is also used on a short-term basis in dangerous manufacturing environments, for example the cleaning and repair of degreasing and chemical plant. For long-term exposure, such as when shot, grit and vapour blasting, an uncontaminated source of air via an airline is provided. Exposed skin should be covered as much as possible to prevent absorption of the contaminant. Radiation intensity can be detected and measured with a Geiger counter.

Physically dangerous environments

Physically dangerous environments include areas containing machinery, fabrication equipment, lifting gear or catwalks. The various types of safety clothing and or equipment worn to protect various parts of the body have already been described and must be selected and worn according to the hazard or hazards likely to be encountered.

Electrically dangerous environments

Most workplaces in manufacturing can be electrically dangerous environments with electric shock being the most common danger. Electric shock is the effect on the nervous and muscular systems of the body caused by an electrical current passing through it. The tolerance of each individual to electric shock varies widely. Shock effects can be produced in some people by electrical potentials as low as 20 volts, and deaths have been recorded by electrical potentials as low as 100 volts but these are very rare.

For normal, healthy persons equipment working from a 110 volt supply is considered safe. Most portable industrial equipment is now designed to run off this voltage via a transformer. An electric shock causes the muscles to contract which in itself may not be serious, but it could cause injury or death if a person is working high up on a ladder, gantry, catwalk or platform and the muscular convulsion causes them to fall. More dangerous is the effect of the shock on the functioning of the heart as this can lead to unconsciousness and death.

The following safety equipment should be used when working with electricity to prevent electric shock:

- Rubber soled shoes or boots.
- Rubber gloves.
- Insulated mats and sheets.
- Insulated tools (hand and power).
- Lightweight aluminium ladders and steps must not be used. Only wooden ladders and steps should be used when working with electrical circuits that cannot be reached from the ground.

Hygienically dangerous environments

Hygienically dangerous environments are those in which workers can contaminate or be contaminated by the products on which they are working. Such environments include food processing areas and pharmaceutical plants. When working with food, the following should be used to ensure that food is not subject to the risk of contamination:

- Head covering such as hats, caps and hairnets, completely enclosing the hair.
- Clean, washable, light coloured protective overalls, preferably without pockets into which particles of contaminants may collect. Protective clothing should be changed and cleaned/laundered regularly.
- Ordinary clothing should be completely enclosed by protective clothing.
- Thin, transparent, plastic gloves used when handling food should be of the disposable type and worn only once.

4.1.14 Safety equipment and systems

Safety equipment and systems include the following:

- guards;
- simultaneous (two handed) control devices;
- trip devices;
- visual and audible warning devices;
- mechanical restraint devices;
- overrun devices;
- safety valves;
- fuses;
- circuit breakers;
- residual current circuit breakers;
- emergency stop buttons;
- limit switches.

Guards

Guards are used to protect operators from the dangerous moving parts of machinery. They may be fixed, in which case they must be tamper proof, or removable in order to adjust or change items, in which case the machine must be isolated from the electrical supply. This is usually achieved by means of a cut-out switch mounted on the guard and connected to the electrical mains supply to the motor so that the machine cannot run when the guard is open. Some guards are adjusted prior to machine operation, such as those mounted on drilling machines, power saws and presses. Other types of guard keep the dangerous part out of reach, such as those fitted to guillotines and cropping machines. Guards must be maintained in good condition by regular inspection and adjustment as necessary. Some examples are shown in Fig. 4.32.

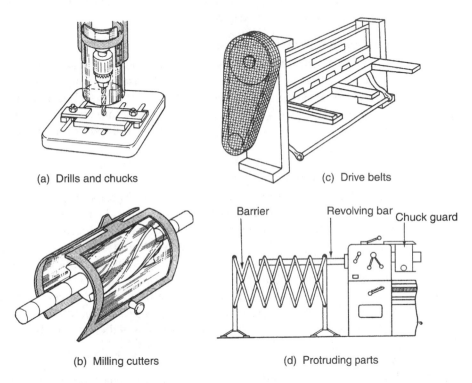

(a) Drills and chucks   (c) Drive belts

(b) Milling cutters   (d) Protruding parts

Figure 4.32 Some typical guards.

Simultaneous control devices

A simultaneous or two handed control device is a device that requires the operator to use both hands at the same time to operate the controls of a machine. In this way both hands of the operator have to be removed from the hazard zone and, therefore, cannot possibly be at risk. This type of device is used on presses and shoe making machinery.

Trip devices

Trip devices stop a machine when any part of the operator's body goes beyond the safe working limit. They may be pressure, mechanically or photoelectrically activated. Pressure sensitive mats are positioned adjacent to dangerous machinery, such as industrial robots, and anyone stepping on to such a mat causes a switch to be activated resulting in the machine being stopped. A light curtain may also be used. Photoelectric cells are positioned around the hazard zone at a safe working distance. Beams of light shine continually onto the photoelectric cells. If any of the light beams are broken, for example by someone passing through one, the machine is automatically stopped. This is a *fail-safe* device since the machine will be stopped if the lights fail for any reason and remove their protection.

Visual and audible warnings

Visual and audible warning devices such lights, two tone sirens and buzzers are used to attract the operators' attention and warn them that something needs their immediate attention.

Mechanical restraint devices

Mechanical restraint devices are provided to stop a machine when it is not working correctly. This type of device is used on the locking system on pressure die-casting and moulding machines. Interlocked guards are fitted to power presses. Here, the machine cannot operate while the tools are being loaded or unloaded. It can only be operated following the operator closing the guard.

Overrun devices

Overrun devices are fitted on machines with interlocking guards. Although the power to the machine is switched off while the guard is open, the momentum of the moving parts prevents the machine from coming to rest immediately. Under these circumstances, the operator may have access to the moving parts. An overrun device delays the opening of the guard and prevents access until the machine has stopped.

Overrun devices of a different type are fitted to lifting gear such as pulley blocks and hoists. This prevents the lifting device going into reverse, allowing the load to descend, when the operator releases the rope or chain.

Safety valves

Safety valves are used to prevent pressure exceeding a preset maximum on boilers and air receivers. They should be tested regularly by a suitably qualified person and the result of the test and its date should be logged. Safety valves should have sufficient capacity so that they can release the steam faster than it can be generated, and release air or hydraulic fluid faster than it can be pumped.

Fuses

Fuses are used to protect electrical circuits from current overload. The fuse fitted should be of the correct type and rating for the appliance or circuit it is to protect. Fuses do not blow at their stated rating. The rated current is the maximum current that the fuse can carry continuously. The fusing current may be as high as 1.5 times or 2 times the rated current. This is called the fusing factor. Fuses only provide *coarse protection*.

Circuit breakers

The circuit breaker is a device that trips a switch to the off position when a current overload occurs. Although it can be easily reset to the on position, it is best to determine the reason for the current overload before doing so. The circuit breaker fitted should be of the correct type and rating for the appliance or circuit it is to protect. Unlike fuses, circuit breakers provide *close protection*. The tripping current is only very slightly greater than the rated current that can be carried continuously. Further the tripping time in the event of an overload is only a matter of a few milliseconds.

Circuit breakers with residual current detection (RCD)

Circuit breakers with residual current detection protect against earth leakage faults. They should be tested regularly to ensure that they are operating correctly. RCD protection is used to protect circuits where a fault causing a current leakage to earth is insufficient to trip the circuit breaker but more than enough to cause an electric shock. The device monitors the current flow in the live and neutral conductors, and if there is any difference, however slight, due to some of the current leaking to earth the contact breaker is tripped immediately. Such devices are used in the supply system for portable power tools especially where they are used out of doors and in damp environments.

Emergency stop buttons

All manufacturing plants with electrically powered machines and equipment will have conveniently positioned emergency stop buttons. When pushed all electrical machinery and equipment in the plant will stop.

The button is bright red and may have a bright yellow surrounding panel, and be sited in prominent positions. This device isolates the plant at the main switchboard, and individual machines or pieces of equipment cannot be restarted by pressing their start buttons. All employees in the plant should be familiar with the position of these buttons. Only an authorized person can switch the electricity back on again.

Test your knowledge 4.12

1. Explain briefly what equipment you would use to deal with:

   (a) a fire in a chip pan;
   (b) a fire in an office stationery cupboard;
   (c) a fire in a photocopier.

2. Explain what emergency action you would take if a workmate's clothing caught fire.

3. Explain the action you would take, and the order in which you would take it, if you found a major outbreak of fire in a hazardous area of your works such as a paint store.

4. State what action you would take if you suffered a minor cut to your hand. State the reasons for your actions.

5. List the protective clothing and equipment you would wear in:

   (a) an engineering machine shop;
   (b) a food processing plant;
   (c) a drop forging shop;
   (d) a woodworking (joinery) shop.

6. Explain the precautions you would take when using portable electric power tools out of doors.

7. State the action you would take if you found that the guard on your machine was not properly adjusted.

4.1.15 Safe working practices

Health and Safety at Work Act

The *Health and Safety at Work Act* provides a comprehensive, integrated system of law dealing with the health, safety and welfare of workpeople and the general public as affected by work activity. Not only are employers and employees *equally responsible* under the Act for ensuring safety at work, so also are manufacturers of equipment and suppliers of goods and materials. Employers are responsible for the health and safety of their employees by the provision and maintenance of safe working conditions. Equally, employees must obey the safe working practices as laid down by the employer, with guidance from the Health and Safety Executive (HSE). These safe working practices are based on the following:

- established and proven practices;
- makers' instructions and guidance notes;
- data sheets;
- health and safety regulations;
- training courses.

Employees must receive supervision and training, especially when new processes or methods are introduced. You must never operate any equipment until you have received training in its correct and safe use. Even then you must not use any equipment without the permission of your supervisor.

Maintenance procedures

The Health and Safety at Work Act makes employers responsible for the safe health risk-free provision and maintenance of plant and systems at work. To meet this obligation, managers must ensure that safety equipment and systems are regularly inspected and maintained in the correct manner at all times by instituting appropriate maintenance procedures, that is, planned maintenance. All manufacturing organizations will have a maintenance section or department, either 'in-house' or 'contracted out', whose function is to carry out these tasks, and will include skilled workers in the fields of mechanical, plant and electrical engineering, and building and building services.

Fire alarms

Fire alarms are used to warn all staff of the danger of a fire or other emergency. All staff should understand the action to be taken when a fire alarm sounds, and evacuate the workplace safely and quickly. Fire instruction notices should be placed at key points around the workplace and in each room. Fire drills should be carried out at least once each year and the fire alarms sound tested regularly. The result of the fire drill (e.g. the time taken to evacuate the premises) should be logged.

Training

All operators need to be fully trained in the safe and correct use of the machines and equipment used in their place of work. They will also need to be retrained when new equipment is installed. Sometimes this training will be carried out 'in-house' but sometimes the operators will need to attend training courses organized by the manufacturers of the equipment or systems. In some instances the operator will need to be independently certificated – for example, only a trained and certificated person may change a grinding wheel.

Supervision

To do their job properly and command the respect of the workers in their charge, supervisors not only require wide experience and knowledge of the process skills involved, they also need to be properly trained in management and interpersonal skills. They should also be competent to train new recruits to their area of responsibility.

Safety equipment and systems

Safety equipment and systems specific to various processes have already been discussed. Further examples of safety equipment and systems will be introduced from time to time in the rest of this chapter.

Protective clothing and equipment

The use of protective clothing and equipment has already been introduced. It is essential to cooperate with the management in using the equipment provided not only to protect yourself but also, in some instances, to render the product free from contamination. Sometimes the protective clothing may be uncomfortable and inconvenient to wear. To neglect its use for such reasons is short-sighted since you not only put yourself at risk of physical injury, you and your employer are both at risk of legal prosecution.

Cutting tools

Even simple hand tools such as scissors, hand shears (tin-snips) and knives can cause a nasty cut if not used correctly. Always carry them point downwards when walking about. Always cut *away* from yourself and *away* from your free hand. Files should be used with both hands to give proper control and must be fitted with a handle in good condition. Scrapers should also be used so that the scraping action is away from your body. Saws should be carefully guided at the beginning of a cut with the minimum of force to prevent it slipping. There is more danger associated with a wood saw than a metal cutting hacksaw since only one hand is used and the teeth are much coarser. Once the cut has been started keep your free hand away from the line of the cut and the teeth of the saw.

The cutters on machine tools should be properly guarded not only to protect your hands but, in the case of drilling machines, to prevent you being scalped. Milling machines and guillotines are particularly dangerous and must not only be fully guarded, they must only be used by fully trained operators.

Assembly tools

Sewing needles should be used with a thimble and when using a sewing machine you should keep your fingers well away from the line of sewing. Soldering irons need to be treated with care as they operate at temperatures sufficient to inflict a nasty burn and also start a fire if laid down on cloth, paper or wood. The blades of screwdrivers should be matched to the head of the screw to reduce the risk of the screwdriver slipping and marking the work and/or damaging the head of the screw. Never hold the workpiece so that the screwdriver is pointing towards your hand. A slip could result in the screwdriver stabbing into your hand.

Spanners and socket screw keys should fit the hexagon head accurately so that they cannot slip. An oversize spanner with packing should never be used. If the spanner slips, you can receive a nasty injury and adjacent components may be damaged. Never extend a spanner as the increased leverage can overstress or break the fastener. Where it is important to achieve the exact tightness specified by the designer, you should use a torque spanner set to the required value.

Test your knowledge 4.13

1. Under the Health and Safety at Work Act, name the groups of persons responsible for safety.

2. Name the organization that implements the Health and Safety at Work Act.

3. Under what conditions is it legally acceptable for you to operate a piece of equipment or a machine at your place training or your place of work?

4. In addition to the time taken to evacuate the premises, suggest other important information that you think should be logged on completion of a fire drill.

5. Find out what is meant by:

   (a) a prohibition order;
   (b) an improvement notice.

6. Explain briefly why operators should be properly trained in the use of manufacturing equipment.

7. Explain how supervisors can contribute to the safety of the workplace for which they are responsible.

Key Notes 4.1

- The *mechanical properties* of materials refer to such properties as strength, toughness, hardness, ductility, plasticity and elasticity.
- The *physical properties* of materials refer to such properties as electrical and thermal conductivity, electrical and thermal insulation, refractoriness, density and magnetic properties.
- *Resistance to degradation and corrosion* refers to the ability of a material to withstand its service environment without deterioration, for example the rusting of steel and the perishing of rubber.
- *Composition* refers to the elements, compounds and mixtures that make up a material.
- *Strength* refers to the ability of a material to resist fracture when subjected to a tensile or a compressive load.
- *Toughness* refers to the ability of a material to resist fracture when subjected to a transverse impact load.
- *Brittleness* is lack of toughness in a material. It is often associated with hard materials.
- *Rigidity* (stiffness) refers to the ability of a material to resist distortion when subjected to an applied load.

- *Elasticity* is the ability of a material to deform when subjected to an applied load and return to its original size and shape when the load is removed.
- *Plasticity* is the ability of a material to deform when subjected to an applied load and to retain its new size and shape when the load is removed.
- *Ductility* is a special case of plasticity when the applied load is of a tensile nature.
- *Malleability* is a special case of plasticity when the applied load is of a compressive nature.
- *Hardness* is the ability of a material to resist indentation and/or scratching by another hard body.
- *Electrical conductivity* is the ability of a material to conduct electricity. A highly conductive material, such as the metal copper, has a low resistance to the flow of an electric current.
- *Electrical insulators* are materials that are very poor conductors of electricity.
- *Thermal conductivity* is the ability of a material to conduct heat energy.
- *Thermal insulators* are materials that resist the conduction of heat energy. Materials that are good conductors or insulators of electricity are also good conductors or insulators of heat energy.
- *Refractory* materials are able to resist high temperatures without softening or melting.
- *Soft magnet materials* such as pure iron (ferrite) only become magnetized when placed in a magnetic flux field. They lose their magnetism when the magnetizing field is removed.
- *Hard magnetic materials* such as quench hardened high carbon steel also become magnetized when placed in a magnetic flux field, but retain their magnetism when the magnetizing field is removed.
- *Density* is the mass per unit volume of a material.
- *Flow rate* is the ease with which a fluid will flow.
- *Flavour* is a property of food and drink determined by taste. It is closely associated with smell.
- *Colour* is the visual effect produced on the eyes by light of differing wavelengths.
- *Composite materials* consist of a matrix of one material reinforced by fibres or particles of another material.
- *Mixtures* are an intimate association of particles of various materials that retain their individual properties and have not reacted together.
- *Compounds* are formed when two or more substances combine together chemically to produce an entirely new substance. Heat has to be taken in or given out for the reaction to take place and the new substance formed has different properties than the substances from which it is made.
- *Alloys* are composed of metals or metals and non-metals in which the atomic particles are arranged in strict geometrical patterns. They have superior properties to pure metals for many purposes. Brass is an alloy of copper and zinc. Steel is an alloy of iron and carbon.
- *Degradation* is the deterioration of non-metallic materials by the effects of ultraviolet light in the case of rubbers and plastics, and the rotting effect of water in the case of wood.
- *Corrosion* is the eating away of metals when subjected to the oxygen of the atmosphere in damp surroundings, for instance the rusting of iron.
- *Erosion* is the physical wearing away of materials by such atmospheric effects as wind and rain.
- *Weathering* is the physical degradation of materials, particularly building materials, due to erosion, pollution, frost, sunlight, etc.
- *Ferrous metals* are those metals and alloys based mainly on iron (Latin: *ferrum* = iron).
- *Non-ferrous metals* are all the metals and alloys that do not contain any appreciable amount of iron or no iron at all.
- *Annealing* is the softening of metals by heating and cooling. The temperature and cooling rate varies according to the type of metal.
- *Normalizing* is similar to annealing but the cooling rate is slightly quicker resulting in a finer grain structure. Normalizing is frequently used as a stress relief process after rough machining and before finish machining in order to prevent subsequent distortion.
- *Work hardening* is the only way many non-ferrous metals can be hardened. When drawn, rolled or pressed while at room temperature their grain structure becomes distorted and this results in hardness.
- *Quench hardening* is a process where medium and high carbon steels are raised to red heat and cooled quickly (quenched) in water or oil. The steel becomes hard and brittle.

- *Tempering* is a process where the quench hardened steel is reheated to about twice or three times the temperature of boiling water and quenched again. This imparts toughness to the steel but with some loss of hardness.
- *Solution treatment* is a heat treatment process where aluminium alloys containing copper can be softened.
- *Precipitation age hardening* is a natural process in which solution treated aluminium alloys start to reharden at room temperature as the copper comes out of solution with the aluminium. The process can be accelerated by heating, or retarded by refrigeration.
- *Compression moulding* is a process for manufacturing articles from thermosetting plastics such as bakelite and melamine.
- *Injection moulding* is a process for the manufacture of components from thermoplastics.
- *Extrusion moulding* is a process for manufacturing lengths of thermoplastic sections such as curtain rail.
- *Blow moulding* is a process for manufacturing bottles from thermoplastic materials by inflation.
- *Vacuum forming* is a process by which heated thermoplastic sheet is moulded over or into formers by atmospheric pressure.
- *Casting* is a process in which molten materials are poured in a mould and left to solidify.
- *Die-casting* is a process in which the molten metal is injected into metal dies.
- *Forging* is a process for forming heated metal by hammering or pressing between shaped die blocks.
- *Sheet metal pressing* is a process where sheets of ductile metal are formed by pressing them into or over shaped dies in a powerful press at room temperature.
- *Galvanizing* is a process in which steel articles are dipped into molten zinc to give them a coating that makes them more corrosion resistant.
- *Conversion coating* is a process where steel articles are dipped into chemical solutions to give them phosphate, oxide, or chromate coatings which not only make them more corrosion resistant but also provide a key for a subsequent painting process.
- *Electroplating* is a process for coating a product with a corrosion resistant protective or decorative metal finish, for example chromium plating.
- *Anodizing* is an electrolytic process for putting a protective and decorative film on aluminium and aluminium alloy products.
- *Glazing* is a process for putting a vitreous (glass-like) finish on ceramic and metal articles. It is also the name give to the finishing of pastry products in the catering industry and the chemical treatment of fabrics, cards and papers to make them shiny.
- *Laminating* is the application of a 'formica' plastic sheet to the chipboard worktops of kitchen units.
- *Bonding* is the joining of materials using various types of adhesive.
- *Soldering* is the joining of metal using a low melting point alloy of tin and lead.
- *Brazing* is a similar process but uses a higher melting point alloy such as silver solder or brass and gives a stronger joint.
- *Welding* produces a joint between to metal components by melting their edges and allowing them to fuse together. Additional 'filler metal' of similar composition to the components is added to prevent thinning of the joint. A very strong joint is made this way. Very high temperatures are involved.
- *Sewing* is the traditional method of joining fabrics.
- *Toxicity* is the property of a material or substance to have a poisonous effect on the body.
- *Narcotics* (in small quantities) induce disorientation, dizziness, drowsiness, giddiness and headaches. In large quantities, they can induce unconsciousness and death.
- *Oxidation* is a chemical reaction with atmospheric oxygen and causes the corrosion of metals, and the combustion of fuels. It can also cause the spoiling of foodstuffs.
- *Flammability* is the ease with which a substance will ignite and burn in the presence of air (oxygen).
- *Perishability* is the ease with which materials will perish, spoil or rot over a period of time.
- *Contamination* is the inadvertent contact or mixing of a required substance with other materials or organisms so that the required substance becomes spoiled.
- *Hygiene* is the maintenance of personal cleanliness and the cleanliness of processing equipment.
- *Sharpness* is the ability of a cutting edge to cut other materials.
- *Discolouration* is the fading or changing of the colour of a material in the presence of sunlight or other causes.

- *Shelf life* is the maximum time a product can be kept in stock without deterioration.
- *Manual (non-automated) machinery* is machinery that is under the control of a human operator.
- *Automated machinery* is machinery that can operate in a repetitious manner without the intervention of a human operator. The control system may be mechanical or by a computer system.
- *Cutting speeds* are calculated using the formula $N = (1000S) \div (\pi \times d)$ where $N =$ the spindle speed in rev/min, $S =$ the cutting speed in m/min, $d =$ the diameter of the workpiece or the cutter.
- *Cutting times* are calculated using the formula $T = (60P) \div (N.F)$ where $T =$ the cutting time in seconds, $P =$ distance of travel of the cutting tool in millimetres, $N =$ the spindle speed in rev/min, $F =$ the feed rate in mm/rev.
- *Cycle time* is the sum of the individual process time elements.
- *Cutting tools* are hand and machine tools used to cut and trim the materials used in manufacture.
- *Shaping tools* are tools used to change the shape of materials without resorting to cutting–there is no loss of volume.
- *Assembly tools* are tools used in the assembly of individual components to make a finished product.
- *Fires* are caused by flammable materials being raised to their ignition temperatures in the presence of air (oxygen). Fires can be put out by removing one or more of the following: source of heat, flammable material (fuel), air (oxygen) fires may be controlled by one or more of the following devices: automatic or dry riser sprinkler system, fire hose, fire blanket, pressurized water extinguisher, foam extinguisher, dry powder extinguisher, CO_2 extinguisher, vaporizing liquid extinguisher.
- *First aid* equipment sufficient for the number of persons employed must be kept in an accessible place and all injuries must be logged. An adequate number of employees trained in first aid must be available.
- *Personal safety clothing* appropriate to the work being done must be made available and must be worn/used by the personnel concerned by law.
- *Hazardous environments* refer to environments that may be dangerously cold, dangerously hot, contaminated, physically dangerous, electrically dangerous, hygienically dangerous.
- *Electrical equipment* must have all exposed metalwork earthed or be double insulated. All electrical circuits must be protected by suitable fuses or circuit breakers. Portable electrical equipment for use out of doors or in a damp environment must be operated via a 110 volt transformer and protected by a residual current detector (RCD) as well as the usual fuses or circuit breakers.
- *Fuses* are devices that prevent an excessive electrical current flowing in a circuit. An excessive current flow could cause the circuit wiring to overheat and start a fire.
- *Circuit breaker*, an alternative, more sensitive, form of electrical circuit protection. See Fuse.
- *Guards* enclosing all moving parts of a machine's transmission system must be fitted by the manufacturer. Guards preventing contact with the cutting tools must also be fitted and are the responsibility of the user of the machine or equipment.
- *Simultaneous control* necessitates the simultaneous use of both hands to set a machine in motion, thus removing both hands from the hazard zone during the operating cycle.
- *Trip devices* automatically disable the machine if persons move into a hazard zone or remove a guard.
- *Visual and audible* warning signals draw attention to equipment malfunctioning.
- *Mechanical restraint* devices disable the machine if it is malfunctioning in a potentially dangerous manner.
- *Overrun devices*. Due to the momentum of its moving parts a machine may not come to a standstill immediately it is switched off. An overrun devise prevents the guards from opening until the machine is stationary.
- *Safety valves* prevent an unsafe build-up of pressure in a boiler, compressed air receiver, or a pressurized system.
- *Emergency stop button*. A large, red coloured, mushroom headed button on the control panel of a machine that disables and stops the machine when pressed in an emergency. It may also refer to similar buttons strategically placed that can disable the supply to all machines in a workshop in the event of an emergency.
- *Health and Safety at Work Act*. Legislation concerned with the health, safety and welfare of employees and the general public who are at risk from any form of manufacturing, or commercial activity. The Act is administered by the Health and Safety Executive (HSE), and responsibility for safety under the Act is shared equally between employers and employees.

Evidence indicator activity 4.1

1. Write a report for a product of your own choosing in *detail*. The report to include a list of the key characteristics of the materials required, a description of the processing methods and a summary of the factors affecting their handling and storage. In each case consider the key characteristics.
2. Write a report for another product of your own choosing in *outline only*. This product should use a contrasting group of materials. The report should also include a list of the key characteristics of the materials required, a description of the processing methods and a summary of the factors affecting their handling and storage. In each case consider the key characteristics.

4.2 Processing materials and components

4.2.1 Safe operation of equipment, machinery and tools

The safe operation of tools, equipment and machinery has been emphasized repeatedly in Sections 4.1.14 and 4.1.15. Whenever an employee or trainee has any doubts as to the correct and safe operation of any tool, piece of equipment or machinery, always enquire who is the competent person to give you advice, and ask that person. Always remember 'IF IN DOUBT ASK'.

4.2.2 Control and adjustment of machinery and equipment

The control and adjustment of machinery and equipment must be carried out in the correct manner in order to process materials and components to specification and required quality. Quality standards are specified by the customer and will include consideration of such items as materials and tolerances in order to meet specification. Let's consider some of the elements that control the overall quality of a product.

- *Linear dimensions*. These are, literally, straight-line dimensions, that is, the shortest distance measured between two points. They are the dimensions to which the product is made. As stated previously, we can neither manufacture a part to an exact size, nor can we measure it exactly. So let's consider Fig. 4.33(a). The designer has given the 25 mm diameter a *tolerance* and has set the *limits of size* for the dimension as ±0.1 mm, that is, any dimension between 24.9 mm and 25.1 mm inclusive is acceptable and the component will function correctly within its quality specification. The tolerance in this instance is 25.1 mm − 24.9 mm = 0.2 mm.
- *Geometric form*. Referring again to Fig. 4.33(a), we see that the 25 mm dimension is a diameter. Therefore this feature of the component must be round (circular). This may be the case even if the dimension is within limits. It could be elliptical in form as shown in Fig. 4.33(b). It would be correct dimensionally and would be incorrect geometrically. Where the form of a component is critical, geometric tolerances as well as linear tolerances must be shown on the drawing. These are rather complicated

and, fortunately, we do not have to concern ourselves with them at this level.

- *Physical and mechanical properties*. The physical and mechanical properties, such as strength, hardness, electrical and thermal conductivity, have already been discussed in some detail, as well as techniques for testing some of these properties. However, quality is also concerned with surface finish. Surfaces that look smooth to the human eye can look like a lunar landscape full of hills and valleys when viewed under a microscope. It is no use the designer specifying close tolerances if the finish is rough. The mean dimension may lie within limits, but the hills and valleys of a rough surface could be outside the limits as shown in Fig. 4.34(a). Therefore a surface finish symbol and a measured value for the surface finish have to be added as shown in Fig. 4.34(b). Surface finishes are measured electronically by drawing a probe over the component surface, amplifying the electrical signal produced and printing it out as a trace on a chart.
- *Materials*. The easiest way to ensure that materials are satisfactory is for the designer to specify materials that conform to a BSI specification and then demanding proof that the batch supplied complies with this specification. For critical components the customer may wish to be present when the material is tested or may sample and retest the material upon delivery.
- *Appearance*. The finish and colour of a component is much more difficult to assess and is largely subjective in that acceptance or rejection depends upon the experience and judgement of the customer. Again, it is best to specify BSI standard colours so that these can be compared with established standards. Surface defects, such as dents and chips, can only be detected by the sharp eyes of the inspector.
- *Sensual effects* – feel, taste, noise level, etc. Noise levels can be measured but sensual effects, such as feel and taste, and can only be assessed by the skill and experience of the inspector.
- *Functionality*. Input level, output level, efficiency, mass, speed or volume can all be measured. Published reports of performance levels are highly important when comparing the suitability of competing products.

Manual equipment and machinery

Manually operated machines need continuous setter or operator attention. The controls have to be adjusted in order to process materials to the desired quality specification. For work where very precise measurements are not necessary, as in clothing manufacture, sheet metalwork and woodwork, patterns and templates may be used for guiding the initial cutting-out operations. Engineering operations are generally of a more precise nature and the operator may work to accurately scribed lines for one-off, prototype components. Where greater accuracy is required, the scribed lines are only a guide and dimensional control relies upon the micrometer dials of the machine controls together with precision measuring instruments such as

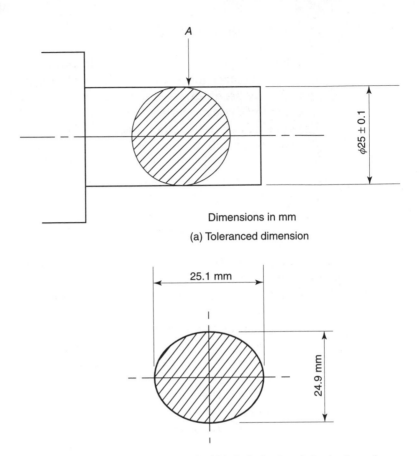

Dimensions in mm

(a) Toleranced dimension

The diameter '*A*' is still within limits but is not circular (round).
It has an incorrect geometric form

(b) Geometric form

Figure 4.33   Dimensions and geometric form.

micrometer and vernier calipers. For batch production, jigs and fixtures may be used. These may:

- position one component relative to another for assembly purposes (welding jig);
- position a component relative to a machine cutter (milling fixture);
- position a component relative to a machine cutter and guide the cutter (drilling jig).

Control and adjustment are not constant, they are only required when the machine produces components that are approaching the dimensional limits or are in danger of drifting outside the desired specification and quality standard. Some machines can be adjusted while they are running (adjusted 'in process'). They may also be adjusted after a pilot run, the final adjustments being made after the product has been inspected and the machine allowed

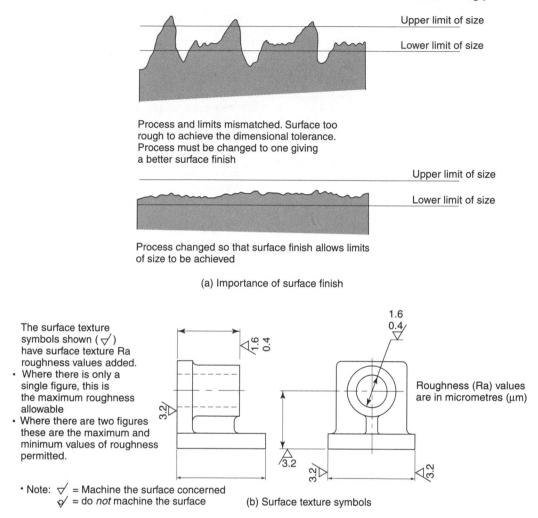

Process and limits mismatched. Surface too rough to achieve the dimensional tolerance. Process must be changed to one giving a better surface finish

Process changed so that surface finish allows limits of size to be achieved

(a) Importance of surface finish

The surface texture symbols shown (∇) have surface texture Ra roughness values added.
- Where there is only a single figure, this is the maximum roughness allowable
- Where there are two figures these are the maximum and minimum values of roughness permitted.

Roughness (Ra) values are in micrometres (µm)

- Note: ∇ = Machine the surface concerned
  ⦶ = do *not* machine the surface

(b) Surface texture symbols

Figure 4.34 Surface texture.

to settle down. Controls that need to be adjusted while the machine or equipment is in operation are positioned so that there is no need to remove guards and no risk to the operator. Such controls are sometimes referred to as 'in-process' controls. During machining or operation a tool may wear and have to be changed and some minor adjustment made as a result. This is a normal part of manual machine or equipment operation.

Typical examples of the parameters that may require adjustment during the operation of manufacturing plant and machinery are shown in Fig. 4.35.

An example of control and adjustment of flow rate is the filling of bottles in a bottling plant. The bottles pass the filling station on a conveyor at a set speed. Therefore the *flow rate* of liquid into the bottles determines the level to which the bottles are filled. If the flow rate is too fast the bottles will contain too much liquid, and if too slow the bottles will not contain enough liquid. Therefore the flow rate is controlled and adjusted as required during the process.

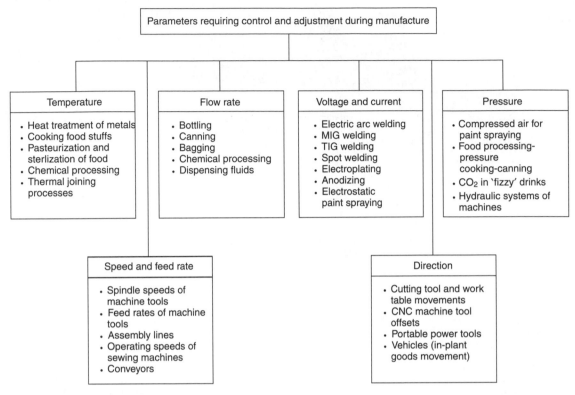

Figure 4.35 Examples of parameters requiring control and adjustment during manufacture.

Automated equipment and machinery

Automated machines are pre-programmed to operate within specified parameters such as speed, temperature, feed rate and flow rate. The operator would need to observe and check that the machine is safe and to specification while carrying out any control and adjustments that might be necessary. Automation can be mechanical where the machine movements are controlled by cams, or pneumatic/hydraulic where the machine movements are controlled by sequencing valves. Computer numerical control (CNC) is widely used these days. The setter/operator programs the machine from the data on the component drawing or may receive the program already prepared on tape or disk. Having loaded the program into the machine controller, the operator then has to adjust the 'offsets' to allow for such variables as cutter length and diameter, lathe tool nose radius, etc. The machine will then produce identical components until the batch is complete or the cutters need to be changed. When the cutters are replaced the 'offsets' will need to be reset but the program remains unchanged. A portion of a program for a CNC milling machine is shown in Fig. 4.36.

Figure 4.37 shows some examples of flow charts for processing typical materials and processes in the manufacturing industries.

It is no good starting up a production line unless the flow of materials and part-finished goods are available to maintain a steady rate of production. If a

Seq No	Code Programme	Explanation
%	GØØ G71 G75 G9Ø	Default line
N1Ø	X – 5Ø.Ø Y – 5Ø.Ø S1000 T1	
N2Ø	MØ6	TC posn spindle tool/offset (06mm)
N3Ø	X25.Ø Y75.Ø Z1.Ø	Rapid to posn (1)
N4Ø	GØ1 Z – 3.Ø F1ØØ	Feed to depth
N5Ø	GØ1 X75.Ø F35Ø	Feed to posn (2)
N6Ø	GØØ Z1.Ø	Rapid tool up
N7Ø	X – 5Ø.Ø Y – 5Ø.Ø S8ØØ T2	
	MØ6	TC posn
N8Ø	X – 5Ø.Ø Y1Ø.Ø Z1.Ø	Rapid to posn (3)
N9Ø	GØ1 Z – 6.Ø F1ØØ	Feed to depth
N1ØØ	GØ1 Y5Ø.Ø F35Ø	Feed to posn (4)
N11Ø	GØØ Z1Ø	Rapid tool up
N12Ø	X – 5Ø.Ø Y – 5Ø.Ø S11ØØ T3	
	MØ6	

Figure 4.36 Portion of a part program for a CNC milling machine BOSS6 (software) (Note: BOSS6 = Bridgeport's own software system, version 6, is the software language for which the program segment shown is written.).

large batch or continuous production is involved, it is also unlikely that there will be sufficient room to take delivery of all the materials in one delivery. Therefore it will be necessary to work out the flow rate per hour and arrive at the amount of material that needs to be available to supply a complete shift without interruption. An example is shown in Fig. 4.38.

Test your knowledge 4.14

1. State the action you should take if you are inadvertently directed to operate a machine that is unfamiliar to you.

2. A high precision component is dimensioned as being 50.15 ± 0.01 mm diameter. Explain why checking this dimension alone will not necessarily be sufficient to ensure that the component conforms to its quality specification.

3. Describe how the duties of an operator using a manually operated machine differ from the duties of an operator using an automatic machine.

4. Draw up a flow chart to show the processing stages for a product of your own choosing.

5. Given the following data, calculate the flow rate of materials required to maintain uninterrupted manufacture.

Key Notes 4.2

- Do not operate any machine or piece of equipment without prior training and, if necessary, appropriate certification of competency.
- Do not operate any machine without permission from your supervisor.
- Do not operate any machine or piece of equipment that you consider unsafe or unfit for service until it has been checked and made safe by a competent person.
- If you are directed to operate a machine that is unfamiliar to you, point this fact out to your supervisor and ask for instruction before you attempt to use it.
- IF IN DOUBT ASK.

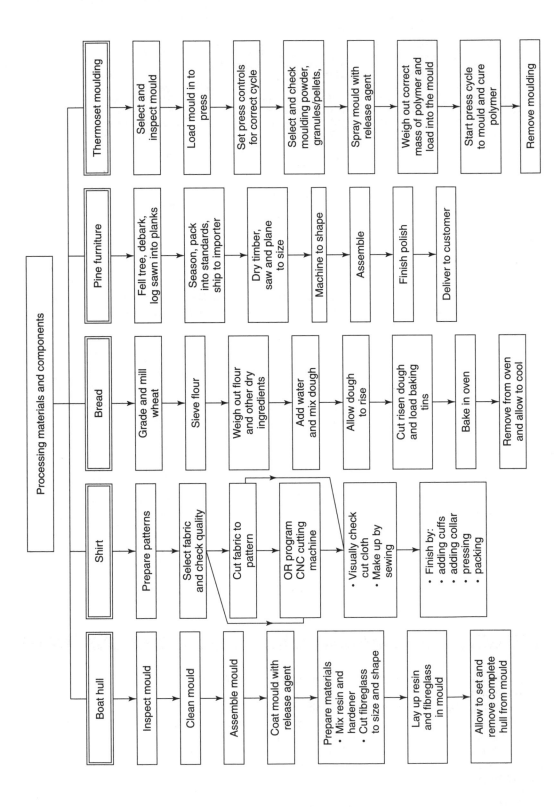

Figure 4.37 Processing meterials.

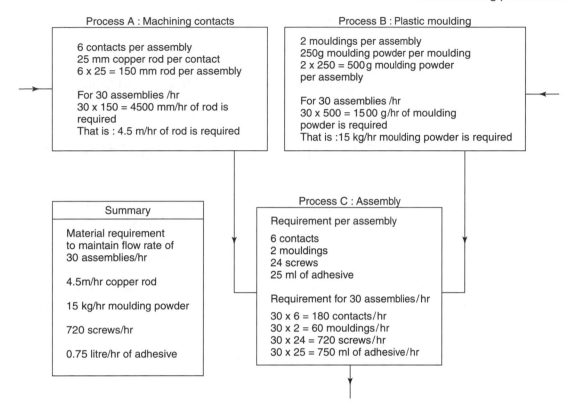

Figure 4.38 Material flow rate to maintain production of 30 assemblies per hour.

- Manually operated machinery and equipment is controlled continuously by the operator all the time it is in operation.
- Automated machines are set up by the operator or setter and will then run unsupervised except for periodic checks to ensure all is well.
- Automation systems may be mechanical (cam operated), pneumatic/hydraulic using sequence valves, or computer numerically controlled.
- Material flow is necessary to maintain an even rate of production without interruption.

Evidence indicator activity 4.2

1. Produce a log briefly recording your performance in practical exercises involving the processing of materials and components to produce a product as directed by your instructor, and which accords to the product specification.
2. Produce a supplementary record of involvement with one other product using contrasting materials.

Note: The log should illustrate your:

- Safe use of equipment and machinery, with reference to manufacturers' instructions.

- Ability to control and adjust the equipment and machinery, with reference to manufacturers' instructions.
- Ability to maintain levels of materials with reference to the specified production flow.
- Knowledge of and ability to check safety equipment, procedures and systems.

4.3 The assembly and finishing of materials to specification

4.3.1 Assembly processes

Assembly is the bringing and joining together of the necessary parts and subassemblies to make a finished product. Sometimes the parts and subassemblies will have been finished before assembly and sometimes after assembly – for instance, the body shell, doors, boot lid and bonnet of a car will have been painted before final assembly. On the other hand, the sheet metal parts and rivets that make up a farm feeding trough will be hot dip galvanized after assembly in order to seal the joints and make them fluid tight as well as sealing the raw edges of the metal against corrosion.

There are three major categories of joining processes used in assembly:

- mechanical;
- thermal;
- chemical.

4.3.2 Mechanical assembly processes

Semi-permanent fastenings

Also called *temporary fastenings*, these include nuts, bolts, screws, locking washers, retaining rings and circlips. They are among the most common methods of joining materials and are used where a joint has to be dismantled and remain so from time to time, for example an inspection cover. When one side of the work is blind, that is inaccessible, or when dissimilar materials are to be joined, self-tapping screws are often used. This type of screw is used to secure such materials as sheet metal, plastics, fabrics, composite materials and leather. Examples of semi-permanent (threaded) fastenings are shown in Fig. 4.39.

Since semi-permanent (temporary) fastenings can be taken apart, they are liable to work loose in service. For critical assemblies such as the controls of vehicles and aircraft this could be disastrous. Therefore locking devices are used to prevent this from happening. Some examples are shown in Fig. 4.40.

Permanent fastenings

Permanent fastenings are those in which one or more of the components involved have to be destroyed to separate the joint – for example, the ring gear shrunk onto the flywheel of a motor car engine has to be split before it can be removed from the flywheel when it needs replacing. At the other end of the scale the cotton or thread used to sew fabrics together has to

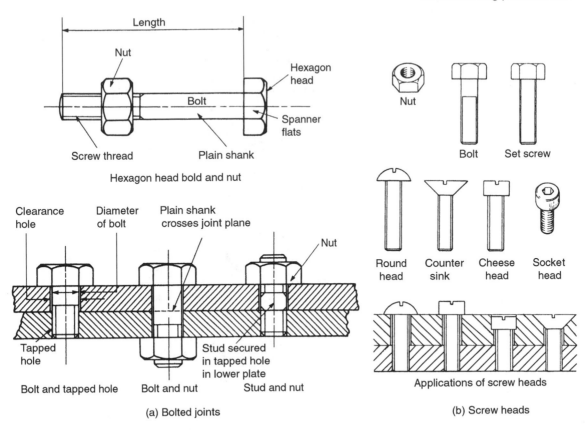

Figure 4.39 Semi-permanent (threaded) fastenings.

be destroyed when a seam is unpicked to make alterations or repairs to a garment.

The major groups of permanent fastenings are riveted joints, metal stitching and compression joints. Rivets are made of ductile materials and depend on deformation to fasten and hold components in place. Rivets and riveted joints have already been introduced in Section 4.1.14. They may be closed either in the red-hot or cold condition depending on the materials to be joined, their thickness and the rivet material. Some techniques for making riveted joints are shown in Fig. 4.41, and some examples of riveted joints are shown in Fig. 4.42.

Metal stitching is a mechanical means of repairing fractures in castings, in which the fracture is prevented from spreading by means of a hole being drilled at each end. A series of holes are then drilled along each side of the fracture and metal clips like staples driven into each pair of opposite holes to seal the fracture. This method of fastening seems implausible but is highly successful and considered a permanent method of fastening.

Compression joints

Compression joints rely upon the elastic properties of the materials being joined. No additional fastenings such as screws or rivets are required. In

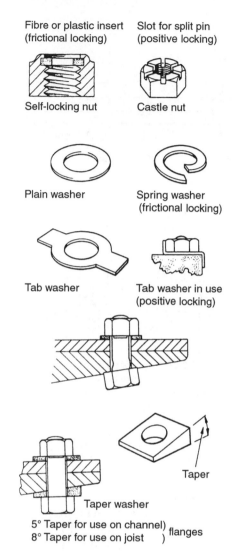

Figure 4.40 Washers and locking devices for threaded fasteners.

principle, a mechanical compression joint, as shown in Fig 4.43(a), consists of a slightly oversize peg being forced into a component with a slightly undersize hole. This results in the peg being compressed and the metal surrounding the hole being stretched. Since the material is elastic, the spring back in the material will grip the peg tightly in the hole where it will be held by friction.

There is a limit to the degree of compression that can be obtained purely by mechanical insertion of the peg into the hole. Where an even greater degree of grip is required two other techniques can be used. In a *hot-shrunk compression joint* the outer component is heated up so that it expands. This enables a much greater difference in size to be accommodated. On cooling, the degree of interference between the components causes one to bite into

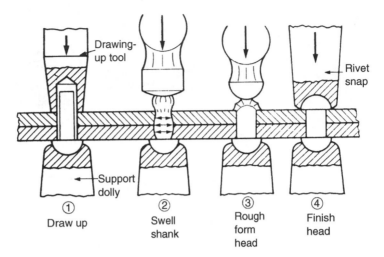

(a) Closing a rivet

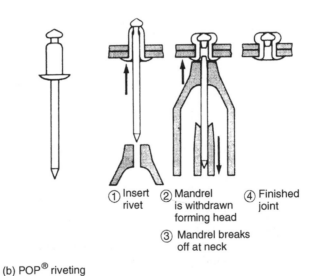

(b) POP® riveting

Figure 4.41 Rivetting processes.

the other resulting in very considerable grip. The principle of this process is shown in Fig. 4.43(b). The starter ring gear of a motor vehicle engine is often shrunk onto the flywheel in this manner.

Unfortunately, when the outer component is heated sufficiently for the required degree of expansion, the temperature may be high enough to alter the properties of the material. The alternative is to use a *cold expansion joint* in which the inner component is refrigerated in liquid nitrogen to cause it to shrink. On warming to room temperature after assembly, the inner component expands and a compression (expansion) joint is again formed as shown in Fig. 4.43(c). Liquid nitrogen can be difficult and dangerous to store and use, so this latter process is only suitable for factory controlled

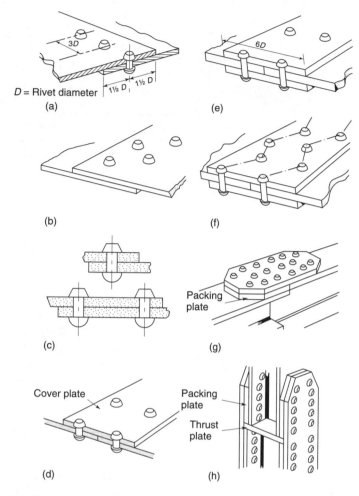

Figure 4.42 Typical riveted joints: (a) single-riveted lap joint, (b) double-riveted lap joint, (c) assembly of lap joints, (d) single-cover-plate butt joint, (e) double-cover-plate butt joint, (f) double-riveted, double-cover-plate butt joint, zigzag formation, (g) splice joint (horizontal), (h) splice joint (vertical).

manufacturing processes. This process does not alter the properties of the component material.

Test your knowledge 4.15

1. State the essential difference between permanent and temporary (semi-permanent) mechanical joints and name an example of each type of joint.

2. Compare the main advantages and disadvantages of mechanical, hot and cold compression joints.

3. One component of a compression joint is made from cast iron, the other is made from steel. State which material would be used for the inner component and which material would be used for the outer component. Give reasons for your choice.

4. Explain briefly why locking devices are sometimes used in conjunction with threaded fasteners and give an example of a typical application.

4.3.3 Thermal assembly processes

The main methods of thermal joining are shown in Fig.4.44.

4.3.4 Fusion welding processes

In all fusion welding processes the filler rod and the edges of the parent metal are melted by a high temperature heat source as shown in Fig. 4.45.

Electric arc-welding

The principle of the *electric arc-welding process* is shown in Fig. 4.46. In this process the heat required to melt the electrode is generated by an electric arc struck between a flux coated consumable electrode, which also provides the filler metal and the workpiece (the parent metal). High temperatures in the range of 4000°C to 6000°C are generated by the electric arc. An electric arc is an elongated and continuous spark. At such high temperatures the metal melts almost instantly in the weld pool. The flux coating of the electrode also melts and some vaporizes. The combination of molten flux and flux vapour protects the weld from atmospheric contamination. On cooling the flux forms a coating over the weld. Arc-welding is mainly used to weld carbon and alloy steels in the fabrication, civil engineering and ship building industries. A heavy current at a low voltage is used for welding. This is supplied by a special welding transformer that has an adjustable output to suit the material and thickness of the workpiece.

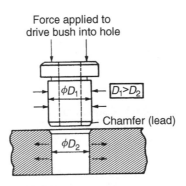

(i) Oversize bush is pressed
into undersize hole

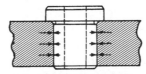

(ii) Spring back of metal
surrounding the hole
grips the bush

(a) Mechanical compression joint

Figure 4.43 Compression joints (permanent).

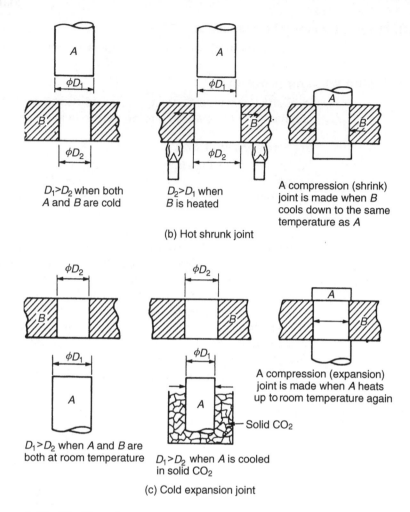

$D_1 > D_2$ when both
A and B are cold

$D_2 > D_1$ when
B is heated

A compression (shrink)
joint is made when B
cools down to the same
temperature as A

(b) Hot shrunk joint

$D_1 > D_2$ when A and B are
both at room temperature

$D_1 > D_2$ when A is cooled
in solid CO_2

A compression (expansion)
joint is made when A heats
up to room temperature again

Solid CO_2

(c) Cold expansion joint

Figure 4.43 (*Continued*).

Two variations of this process that are widely used are:

- Metal-arc Inert Gas (MIG) shielded welding is a semi-automatic process. It is mainly used for welding sheet metal and thin plate in such materials as carbon and alloy steels, aluminium and aluminium alloys, and magnesium alloys in the automobile, petrochemical and light and medium fabrication industries. The entire arc area is shielded by an inert gas that protects the molten pool/electrode and resultant weld from atmospheric contamination.
- Tungsten-arc Inert Gas (TIG) shielded welding uses a non-consumable tungsten electrode to generate an arc between the workpiece and electrode. The molten weld pool and electrode are surrounded by a shielding gas which protects them from atmospheric contamination. A separate filler wire is fed by hand in a similar manner to that employed in the oxy-acetylene gas welding process. The resultant weld is very clean and of high purity. A great advantage of this process is the absence of a

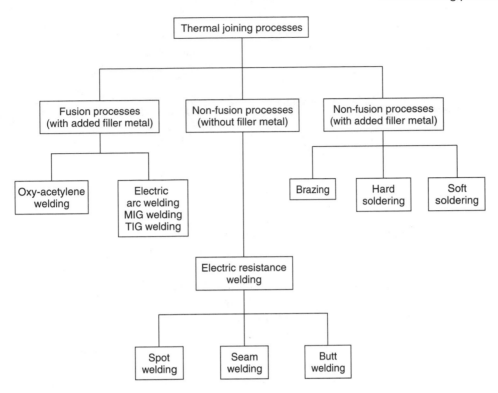

Figure 4.44 Typical thermal joining processes.

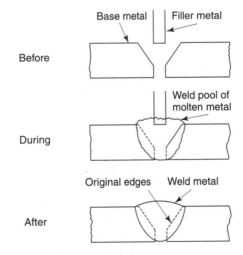

Figure 4.45 Principle of fusion welding with a filler rod.

chemical flux which is generally needed when fusion welding materials with high refractory oxides, for example aluminium and stainless steels. TIG welding is used to weld a wide range of metals and alloys in the petrochemical, aerospace, food and nuclear industries, and on racing car components.

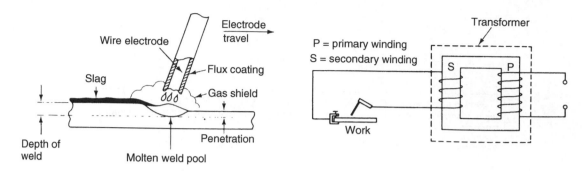

Figure 4.46 Electric arc welding.

Oxy-acetylene (gas) welding

Oxy-acetylene welding is a process in which the heat for welding is produced by burning approximately equal amounts of oxygen and acetylene. The gas is stored in cylinders and is fed via gas regulating valves to a blowtorch. The welding flame produces a temperature of approximately 3500°C and this is sufficient to melt the filler rod and the edges of the metals to be joined. Oxy-acetylene welding is a manual process widely used in the engineering, fabrication and automobile industries on metals such as carbon and alloy steels, cast iron, aluminium and aluminium alloys, brasses and bronzes. The gas is very expensive compared with electricity and the heat energy available is limited. Therefore gas welding is limited to sheet metal and thin plate work. The equipment and principle of oxy-acetylene welding is shown in Fig. 4.47. A flux is rarely needed when oxy-acetylene welding since *the products of combustion* of the oxy-acetylene flame form a blanket over the weld pool and prevent atmospheric oxidation.

Non-fusion processes

In non-fusion thermal joining processes only the filler material becomes molten. The materials being joined remain solid but must be raised to a temperature where a reaction takes place bonding the filler material to them.

Brazing

In the brazing process the filler material is a non-ferrous alloy such as brass (hence the name of the process). Brass filler materials are often referred to as *spelter*. The heat source is a gas torch burning a mixture of oxygen and a fuel gas such as acetylene or propane. The melting temperature of the filler material is above 700°C, but less than the metals to be joined. In all brazing and soldering processes cleanliness of the joint materials is essential. A chemical flux is used, the type being dependent upon the materials to be joined. The entire joint area must be fluxed prior to the application of heat. Once the desired temperature has been reached, the filler rod melts and flows, being drawn between the metals to be joined by capillary attraction to form a strong, durable joint. One advantage of this process is that dissimilar metals may be joined, for example steel tubes to malleable iron fittings when making cycle frames. The process is widely used on a wide and

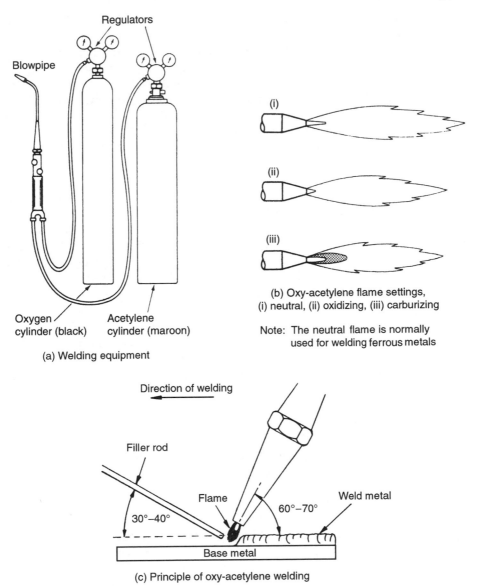

(b) Oxy-acetylene flame settings,
(i) neutral, (ii) oxidizing, (iii) carburizing

Note: The neutral flame is normally
used for welding ferrous metals

(a) Welding equipment

(c) Principle of oxy-acetylene welding

Figure 4.47   Oxy-acetylene welding (gas welding).

varied range of work, on metals such as carbon steel, copper, cast iron and
aluminium.

Soldering (hard)

Hard soldering is similar to brazing except that the filler material contains
the precious metal silver. For this reason the process is often referred to as
silver soldering. Hard soldering is carried out at temperatures above 600°C,
the heat for the process being provided by a welding torch or blowtorch
depending on the size and intricacy of the component to be joined. Since the

solder contains a high percentage of the precious metal silver, it is relatively expensive so that it can only be used economically for the finest work. It produces a neat, strong and ductile joint. A flux has to be used and its composition depends upon the composition of the solder being used. The flux is supplied by the manufacturer of the solder. The materials that can be joined are steel, copper, tin bronze and brass, and some precious metal alloys in similar or dissimilar combinations on light fabrications and jewellery. The solder can be supplied with melting points in 5°C steps to allow a number of components to be sequence soldered onto the same assembly.

Soft soldering

The soft soldering process is carried out at temperatures below 250°C. The solder is essentially an alloy of tin and lead. The heat for the process is provided by a gas or electrically heated soldering iron in the manual process. The solder is supplied in the form of a stick or a wire. For production and automated welding some form of hot plate, resistance heating, oven or induction heating is used, the solder being in the form of a paste, foil or fine wire preform. For electrical and electronic wiring, the solder is often hollow and has a flux core.

As for all brazing and soldering processes, the area to be joined must be physically and chemically clean. This is why the joint must be fluxed prior to soldering, and the iron cleaned, fluxed and tinned before use. Soldering is used to join light sheet metal fabrications, electrical wiring and decorative goods. Fluxes may be *active*, in which case they chemically clean the joint but leave an acid residue that must be washed off after processing. This is unsuitable for electronic wiring or for food canning where a *passive* flux must be used. This will only prevent oxidation and has no chemical cleaning action. In the case of food canning the flux must also be non-toxic. Figure 4.48 shows a typical manual soft soldering operation.

Friction welding

Friction welding is the process of joining metals or plastic materials by rotating one part against the other thereby generating heat by friction until the welding temperature is reached, upon which the two parts to be joined are pressed together until the joint area cools and the joint is made. In addition to similar metals the dissimilar metal combinations that can be joined include steel to aluminium, steel to copper and steel to cast iron. Thermoplastics such as PVC are the only type that can be joined by this method.

Resistance (spot and seam) welding

This is not a fusion welding process. The welding temperature is slightly below the melting point of the metal and the weld is completed by the application of pressure. It is similar in principle to a blacksmith's forge weld.

In resistance welding processes an electric current is passed between two pieces of sheet metal at the point where they are clamped together by the two electrodes. The resistance of the metal to the flow of current causes the metal to heat up locally to its pressure welding temperature. At this point the electric current is turned off and the joint is squeezed tightly between the two electrodes and a spot weld is formed. The cycle of operations is controlled automatically.

Spot welding, as the name implies, makes a weld at the point or spot of contact of the electrode with the sheet metal. This is shown in Fig. 4.49(a). Spot welding is widely used in the manufacture of sheet metal fabrications in the automobile industry.

Resistance seam welding employs two power driven wheels as the electrodes, between which the sheet metal to be joined is passed. The wheels press down hard on the components being joined. A pulsed electric current is passed through the wheels as they rotate. This heats the metal and the pressure of the wheel load makes the weld. In fact the joint consists of a series of overlapping spot welds as shown in Fig. 4.49(b). Resistance seam welding in widely used to make fluid tight joints in the canning industry, in the production of aerosol cans and the production of road vehicle fuel tanks.

Ultrasonic welding

Ultrasonic sound waves are at too high a pitch for the human ear to hear. They can be focused and the high speed vibrations they produce in the materials being joined create a build-up of heat energy enabling a weld to be made when pressure is applied. The process is similar to spot welding but can be used on plastic materials that will not conduct electricity. Instead of electrodes the work is gripped between the ultrasound resonators.

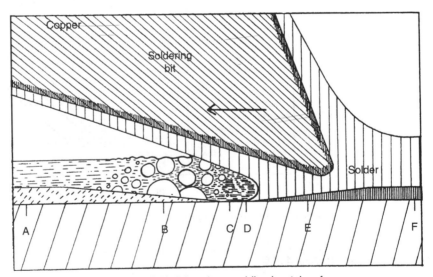

A flux solution lying above oxidized metal surface
B boiling flux solution removing the film of oxide
C bare metal in contact with fused flux
D liquid solder displacing fused flux
E tin reacting with the basis metal to form compound
F solder solidifying

(a) Principle of soft-soldering (courtesy: Tin Reserch Association

Figure 4.48   Soft soldering.

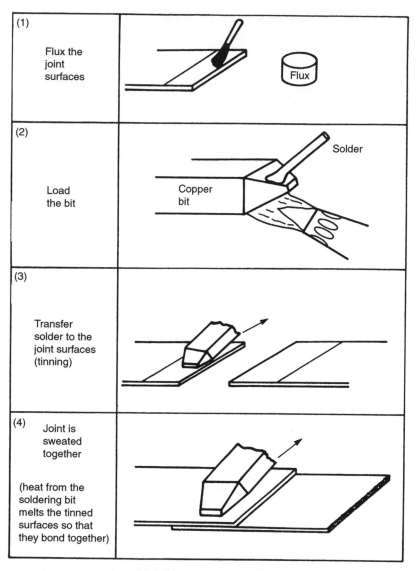

(1) Flux the joint surfaces	
(2) Load the bit	
(3) Transfer solder to the joint surfaces (tinning)	
(4) Joint is sweated together (heat from the soldering bit melts the tinned surfaces so that they bond together)	

(b) Soft soldering a lap joint

Figure 4.48 *(Continued)*.

Test your knowledge 4.16

1. Name the process you would use for the following applications, giving reasons for your choice:

   (a) Welding 25 mm thick steel plate.
   (b) Connecting electronic components to a circuit board.
   (c) Building a copper boiler for a model steam engine.
   (d) Assembling small sheet steel components together.
   (e) Welding sheet metal ducting on site.

2. Explain why ultrasonic welding is used in preference to spot welding when joining plastic materials.

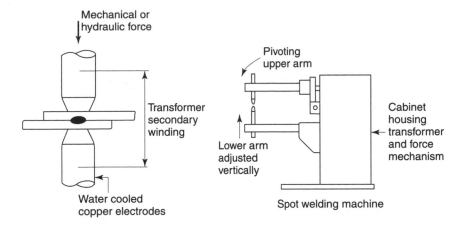

(a) Spot welding

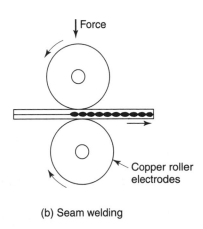

(b) Seam welding

Figure 4.49 Resistance welding processes.

4.3.5 Chemical joining processes

The relationship between some joining processes, depending upon chemical reactions, is shown in Fig. 4.50.

Solvent welding

Solvent welding is the process of joining certain thermoplastics by the application of a solvent to the surfaces to be joined. Polystyrene and poly-methylmethacrylate (acrylic) are the two most widely used materials in this process. There are many solvents that can be used for solvent welding and also numerous proprietary solvent adhesives available. The solvent is applied by means of a brush or pad, the two faces being brought together immediately after application and allowed to dry for a sufficient period of time to ensure adequate joint strength. Solvent welding cements are in this group – for example, the balsa cement used in the building of model

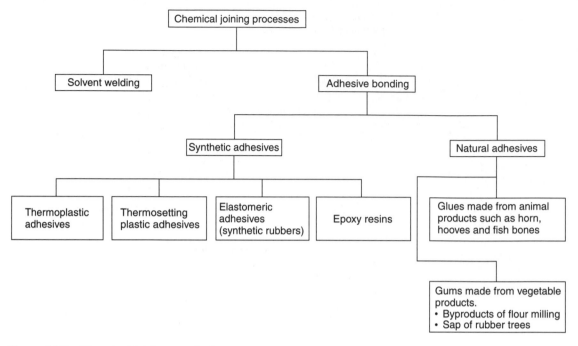

Figure 4.50 Chemical joining products.

aeroplanes. This cement consists of cellulose material dissolved in an acetone solvent. It has 'gap filling' properties and sets by the evaporation of the solvent.

Adhesive (chemical) jointing

Adhesive jointing is used extensively in industry today. There are many different types of adhesives available and they come in many different forms. It is essential to choose the correct adhesive for each application and the manufacturers' literature should be consulted. No matter what the application or the type of adhesive used it is essential that the surfaces being joined are clean and that the humidity and temperature of the workshop is correct. Adhesives may be applied by brush, tube, gun and dipping. The main types of adhesives available are:

- natural;
- thermoplastic;
- thermosetting plastic;
- elastomeric (rubber based).

Natural adhesives

Natural adhesives are of vegetable or animal origin and the simplest in common use. Gums are derived from vegetable matter and include starches, dextrin and latex products. Their use is generally limited to paper, card and foils. Being non-toxic they are safe for use in the food industry for labelling and packaging, and for products where the adhesive is licked, e.g. envelope flaps and stamps. Latex adhesives are used for self-sealing envelopes. Animal glues that are produced from hide, bone, horn and hooves have higher joint strength and were traditionally mainly used in woodworking. They had to be softened by heating over boiling water and set again on cooling (a natural form of thermoplastic). Natural glues have almost been superseded by synthetic glues that are stronger and less affected by their service environment.

Thermoplastic adhesives

These adhesives soften when heated and are subject to creep when under stress, but have good resistance to moisture and biodeterioration. They are used for low loaded and stressed assemblies of wood, metal and plastics.

Thermosetting plastic adhesives

Thermosetting adhesives cure through the action of heat and/or chemical reaction. Unlike thermoplastic adhesives they cannot be softened. These 'two-pack' adhesives consist of the resin and the hardener that have to be kept separated until they are mixed immediately before use. Once mixed the 'curing' reaction commences and there is only a limited time during which the adhesive can be worked. Such adhesives provide high strength joints that are creep resistant and have a high peel strength. However, they tend to be brittle and have low impact strength. Therefore they are unsuitable for joining flexible materials. Epoxy adhesives are in this group.

Elastomeric adhesives

These types of adhesive are based on natural and synthetic rubbers. In general these have relatively low strengths but high flexibility and are used to bond paper, rubber and fabrics. Neoprene based adhesives are superior to other rubber adhesive in terms of rapid bonding, strength and heat resistance. Impact or contact adhesives are in this group.

Test your knowledge 4.17

1. Name the type adhesive you would use for the following applications giving the reasons for your choice:

   (a) the application of 'formica' sheet to the chipboard top of kitchen units;
   (b) assembling the polystyrene components of a model aeroplane kit;
   (c) labelling food products;
   (d) assembling a coffee table;
   (e) resoling a shoe.

2. Describe the essential difference between solvent welding and using an adhesive.

4.3.6 Assembly of components and subassemblies to specification and quality standards

Components and subassemblies must be assembled to the given specification and quality standards to ensure customer satisfaction and confidence in the product. For a simple assembly the general arrangement (GA) drawing is adequate. It shows the relationship between all the component parts, the methods of joining them together and the number of parts required. It does not show the order in which the parts should be assembled.

For more complex assemblies and when moving parts or mechanisms are involved, more detailed and comprehensive assembly drawings are required in addition to the general arrangement drawing. Assembly drawings often incorporate exploded views that indicate the relative positions of the various components and the order in which they should be assembled. In addition to such drawings a schedule of the type and quantities of the components required would also be issued together with a schedule of any assembly jigs and/or special tools that are required.

The process of assembly is the fitting of component parts together to produce either a complete product or a subassembly of a larger and more complex assembly as shown in Fig. 4.51. A typical example of this process is the production of a motor car in which the subframe, engine and transmission, steering mechanism, body shell, doors, and seats are all subassemblies that, on final assembly, result in the finished car.

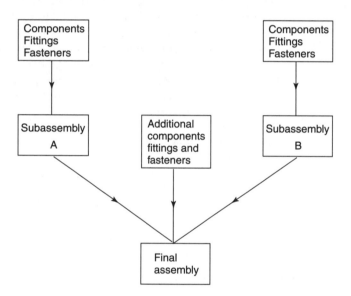

Figure 4.51 Assembly of components and subassembles.

The type of assembly process will depend upon the type of product and the number being produced – for example, motor cars are built in very large numbers on production lines involving the use of industrial robots for some assembly processes and assembly line operatives where only human skills can produce the required result. On the other hand, ships and bridges are built as one-off projects almost entirely by hand except for the use of mechanical lifting and handling aids because of the size and weight of the parts.

A bottling plant or a food canning plant can also use automated conveyor line assembly techniques because of the volume of products made and the repetitious nature of the process. A clothing factory is more likely to work on a batch basis. The ready cut cloth is stacked by the machinists who make up the garments and let them fall into baskets as they are completed ready to be wheeled away to the next workstation.

No matter what method of assembly is used, the quality standards set out in the specification must be achieved – for example, this can involve random sampling in the case of canned foods or bottled drinks. In the case of assemblies with moving parts such as machine tools each machine will not only be rigorously inspected for dimensional accuracy, it will be operated over its full range of feeds and speeds to test for smooth running. Most likely a test piece will be machined to ensure all is well before delivery to the customer.

4.3.7 Finishing methods

Finishing is the final process in the manufacture of a product and as such is very important particularly with regard to product quality. Finishing methods may range from lacquering a polished metal product to wrapping and labelling a food product, and from pressing a shirt prior to packaging to sharpening a knife.

The reasons for applying some sort of finish to a product include providing:

- environmental protection against corrosion and degradation;
- resistance to surface damage by wear, erosion and mistreatment;
- surface decoration.

In general, product finishes should have the following properties:

- they should be colourfast;
- they should provide an even covering of uniform colour and texture;
- they should not run, blister or peel;
- they should be able to resist all environmental and operating conditions;
- they should be resistant to physical damage (e.g. erosion and scratching);
- they should be resistant to staining;
- they should be easy to keep clean.

Different finishes are applied to different products to suit their application and market, for example a metal product may be electroplated to improve its corrosion resistance and/or appearance. On the other hand, a fabric may require a finishing treatment to render it water proof or crease resistant. In some cases it may be necessary to apply more than one finishing process to a product. Furnishing fabrics used for upholstery may need to be finished to render them both stain resistant and flame resistant.

The range of finishes used in manufacturing is very wide and as we have already seen the finishes vary according to the product and the materials used. One way to consider finishes is to group them according to the group of materials to which they will be applied. Some examples are shown in Fig. 4.52.

Another way to group finishes is as follows:

- applied liquids (organic);
- applied liquids (inorganic);

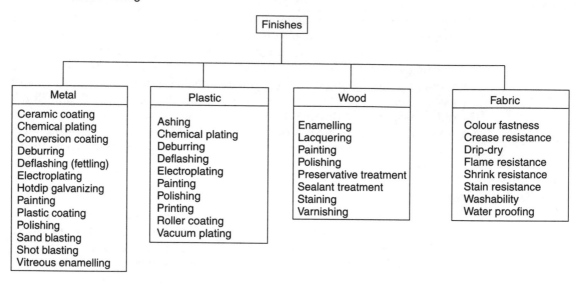

Figure 4.52 Finishes.

- coatings;
- heat treatment processes.

Let's now look at each of these categories in turn

Applied liquid finishes (organic)

Organic liquid finishes include paints, enamels, lacquers and varnishes, etc. Paint consists of pigments (colourants) suspended in a liquid base that can set hard once it has been applied. Traditionally, linseed oil was used because this natural oil 'cures' and sets hard in the presence of atmospheric oxygen. Nowadays synthetic liquids are used that set quicker and harder, for example polyurethane. A complete paint system usually consists of three coats, a priming coat to aid adhesion and prevent rotting or corrosion, an undercoat to build up the colour and a top coat containing a varnish to seal the system and retard or prevent the absorption of moisture. Organic coatings form a decorative, durable and protective coating on the metal or wood to which they are applied. Paints, lacquers and varnishes may be applied in a variety of ways. Some of the more common used in manufacturing industries are:

- brushing;
- dipping and stoving;
- spraying;
- roller coating;
- silk screen printing.

Applied liquid finishes (inorganic)

Compared with organic finishes, inorganic finishes are harder, have increased rigidity and are more resistant to the high temperatures used in cooking. In manufacturing, two of the most widely used groups of inorganic finishes are porcelain enamels and ceramic coatings.

Porcelain and vitreous enamels

These are based on ceramic materials and metallic oxide, metallic sulphide and metallic carbonate pigments. The metal products are coated with a slurry of these materials and then fired in an kiln. The resulting finish is hard, abrasion resistant and suitable for ovenware. The high gloss surface is easy to keep clean and hygienic, hence its popularity for cooking utensils. Unfortunately these finishes are very brittle compared with ordinary paints and varnishes and are easily chipped.

Ceramic coatings

These are hard refractory coating materials, used to protect metal components from the effects of heat corrosion and particle erosion. They are used in such products as:

- gas turbine blades;
- rocket motor nozzles;
- chemical processing plant;
- metal processing plant;
- textile processing plant;
- thermal printers for data processing.

Coating (metallic)

Metallic coating is the plating of a metal product with another metal that has better corrosion resistant and/or decorative properties, thus providing improved:

- surface protection;
- wear resistance;
- decoration;
- dimensional control (gauges are often hard chrome plated to make them more wear resistant – hard chrome plating is used to build up worn gauges so that they can be reground to size).

In the case of hot dip galvanizing (zinc coating) and electrolytic galvanizing the process is often used as a basis for painting. This prevents any moisture that penetrates the paint film from attacking and corroding the metal under the paint film. Corrosion occurring under the paint film, causing it to bubble and lift, was a common problem with cars up to a decade ago.

Metallic coatings may be applied by a number of processes such as electroplating, galvanizing and sherardizing (see Section 4.1.9). Where the product is to be finished by painting, the cheaper sherardizing process may be used instead of galvanizing as a preliminary rust proofing treatment. In the sherardizing process the components are chemically cleaned and heated in contact with zinc powder below the melting point of the zinc. It is a cheaper process than galvanizing but can only be considered as a preparatory process for finishing steel products ready for painting.

Coating (conversion)

Conversion coating is the result of a chemical reaction produced when a corrosion resistant film is deposited on a base metal. The main conversion coating processes are:

- chromating;
- phosphating;
- anodizing;
- oxide coating.

Chromating

Chromating is applied to magnesium alloy components that are dipped into a hot solution of potassium dichromate to form a hard protective oxide coating. The surface is then sealed with a zinc chromate paint and, if necessary, a decorative paint. This process should not be confused with chromizing, which is similar to sherardizing except that the components are heated in contact with chromium metal powder.

Zinc chromate paint is an excellent anti-corrosion primer for use on steel components and structural steelwork prior to painting. A thin coating of zinc chromate primer can be applied to the steel components or steelwork by immersion, spray, brush or roller techniques.

Phosphating

Phosphate coatings are used on refrigerators, freezers, air conditioning units, car bodies and metal furniture to give a high resistance to humidity and weathering. The phosphating chemicals convert the natural oxides on the surface of ferrous metal into complex phosphates that are corrosion resistant. The chemicals are applied by immersion or spray processes. Phosphating is a pretreatment to improve the corrosion resistance of metal components that are finished by varnishing, painting or lacquering. Phosphating is also known as Parkerizing, Bonderizing, Granodizing and Walterizing.

Anodizing

Anodizing is an electrochemical treatment applied to aluminium, magnesium and their alloys. It builds up and hardens the naturally protective oxides on the surface of these metals. The colour of the finish can be controlled to some extent by the chemical composition of the electrolyte used (see Section 4.1.9) or the oxide film can be dyed after processing. It is applied to architectural, aircraft and automobile components, furniture and kitchen fittings, sports equipment, food processing equipment and fashion jewellery.

Oxide coating

Oxide coatings are mainly applied for decoration, but they do provide a degree of corrosion and wear resistance. They are often applied to carbon steel by dipping the heated component in oil which results in a blue/brown, mottled finish.

Heat treatments

Heat treatment processes such as quench hardening and tempering are used for finishing cutting tools and were dealt with in detail in Section 4.1.7.

Annealing

Annealing as a finishing process is applied to metal products to remove stresses caused by previous processing. In a similar process in the textile industry thermoplastic polyester fibre fabrics are heated in the pressed or pleated state which has the effect of relaxing the production stresses in the fibres, causing the pleats to retain their shape upon cooling. It is also used to remove stresses in glass products to render them less brittle.

Test your knowledge 4.18

1. Explain briefly the difference between a general arrangement drawing and an assembly drawing, state where each would be used.

2. For a product of your own choosing, describe the final inspection it would receive to ensure it had met the quality specification.

3. State the finishes that would be used on the following products and the reason such finishes would be used:

   (a) a car body panel;
   (b) an aluminium window frame;
   (c) protective clothing for firemen;
   (d) fabric for cushion covers;
   (e) the hob of a domestic cooker;
   (f) the steelwork of a bridge;
   (g) a sheet steel watering can;
   (h) a garden ornament made from wood (windmill).

4.3.8 Safety when assembling and finishing

Let's now consider the safety hazards associated with assembling and finishing. There are few such hazards associated with mechanical joints. Accidents can occur when using spanners if they do not fit properly but these are usually no more serious than bruised and barked knuckles. Rather more care is required with compression joints. For hot joints, burns are the most likely injury and suitable protective clothing such as heat resistant (*not* plastic) gloves and goggles should be worn. For cold joints involving the use of liquid nitrogen, thermally insulated gloves should be worn together with a full face visor.

There are rather more hazards associated with thermal joining processes and these will now be considered in rather more detail.

Soft soldering

The temperatures associated with soft soldering are relatively low and burns and fire risks are relatively low if the equipment is handled carefully.

Silver soldering and brazing

The temperatures associated with these processes are very much higher. Gloves and eye protection should be worn and tongs will be required to handle the work. If bottled gas torches are used, then the rules concerning the storage and use of bottled gases must be applied. These will be considered under gas welding.

Gas welding (oxy-acetylene welding)

Never use electric or gas welding equipment unless you are under close supervision or have been properly trained. The hazards associated with oxy-acetylene welding can be considered under three main headings:

- eye injuries;
- burns and fire hazards;
- explosions.

Eye injuries

Eye injuries can be caused by the glare and radiated heat of the incandescent (white hot) metal of the weld pool. Also there is a risk of 'splatter' from molten droplets of metal. Therefore, to prevent eye injuries, proper welding goggles must be worn as shown in Fig. 4.53. Ordinary goggles as worn when machining are not suitable since they do not filter out the harmful radiations. Welding goggles have special filter glasses that can be changed according to the metals being joined. In front of the filter glasses are toughened clear glass screens to protect the filter glass from damage. Note that gas welding goggles are unsuitable for electric arc-welding.

Burns

Burns and fire hazards can result from careless use of the welding torch, from sparks and droplets of molten metal and from the careless handling of hot metal. It is important that all rubbish and flammable materials are removed from the working area before welding is commenced. Fire resistant overalls should be worn fastened to the neck. Avoid cuffs as they present traps for sparks and globules of hot metal. A leather apron should be worn. Gloves may be necessary in very hot conditions.

Explosions

The compressed gases and gas mixtures associated with oxy-acetylene welding can cause serious explosions if not properly stored and used. Explosions can occur if acetylene gas is present in air in any concentration between 2 and 82%. It will also explode spontaneously at high pressure without oxygen or air being present. The working pressure for acetylene must not exceed 620 millibars. For this reason the acetylene is not compressed like other gases, but is absorbed in acetone and then stored under pressure in the porous filling of the cylinder. Explosions in the equipment itself can result from improper use, incorrect setting of the equipment and incorrect lighting-up procedures. For these reasons *backflash eliminators* must be fitted and inspected regularly. Remember that an exploding acetylene cylinder is equivalent to a large bomb. It can demolish a building and kill the occupants. Great care must be taken in its use.

Gas cylinders themselves are not dangerous as they are regularly tested to government standards. Nevertheless the following safety precautions must be followed:

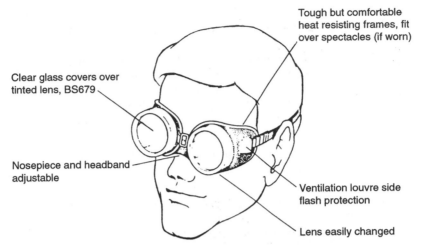

Tough but comfortable heat resisting frames, fit over spectacles (if worn)

Clear glass covers over tinted lens, BS679

Nosepiece and headband adjustable

Ventilation louvre side flash protection

Lens easily changed

Note: Goggles with lenses specified for use when gas welding or cutting must not be used for arc welding operations

(a) Oxy-acetylene welding goggles

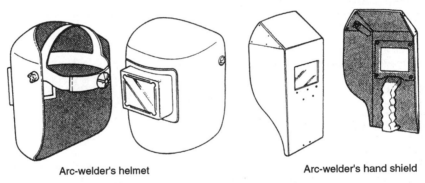

Arc-welder's helmet

Arc-welder's hand shield

(b) Arc-welding eye and head protection

Figure 4.53 Eye protection when welding. Part (a) from *General Engineering* by R. L. Timings, Longman. Reprinted by permission of Pearson Education Ltd.

- Cylinders and their valves must be protected from mechanical damage when being moved and stored. *Acetylene cylinders must always be kept upright when in use.*
- Cylinders must be kept cool. The welding flame must never come into contact with them nor should the cylinders be stored near heaters or in strong sunlight. They must also be protected from frost as this can embrittle and weaken the metal from which they are made.
- Cylinders must be kept in well-ventilated surroundings to avoid an explosive build-up of gas mixtures. Also, oils, greases and clothing can ignite spontaneously if oxygen enrichment of the atmosphere occurs. As stated previously, as little as 2 per cent acetylene in air forms an explosive mixture.
- Automatic pressure regulator valves must be fitted to the cylinders. To prevent them being fitted incorrectly the threads are different for oxygen

and acetylene fittings. The cylinder valve must always be kept closed when the equipment is not in use or when cylinders are being changed.

These notes are the barest outline of the safety precautions that need to be kept. BOC Gases issues a number of booklets on the correct and safe use of gas welding equipment.

When silver soldering or brazing, oxy-acetylene equipment may be used for fine work as the flame can be localized. For general brazing and silver soldering oxy-propane is often used as the process does not require such high temperatures and propane is much cheaper and safer than acetylene. Propane is a liquid petroleum gas. However, the correct regulators for this type of equipment must be used and the hazards associated with oxygen enrichment of the atmosphere are still present.

Arc-welding hazards

The hazards associated with arc-welding are summarized in Table 4.7.

Table 4.7 Arc-welding electrical hazards

CIRCUIT – HIGH VOLTAGE – Primary	
Fault:	Hazard:
1. Damaged insulation	Fire – loss of life and damage to property
	Shock – severe burns and loss of life
2. Oversize fuses	Overheating – damage to equipment and fire
3. Lack of adequate earthing	Shock – if fault develops – severe burns and loss of life
CIRCUIT – LOW VOLTAGE – Secondary (very heavy current)	
Fault:	Hazard:
1. Lack of welding earth	Shock – if a fault develops – severe burns and loss of life
2. Welding cable – damaged insulation	Local arcing between cable and any adjacent metalwork at earth potential causing fire
3. Welding cable – inadequate capacity	Overheating leading to damaged insulation and fire
4. Inadequate connections	Overheating – severe burns – fire
5. Inadequate return path	Current leakage through surrounding metalwork – overheating – fire

The following precautions should be taken to avoid these hazards. These are the basic precautions. Additional precautions may be necessary under some working conditions.

- The equipment must be fed from the mains supply via an isolator incorporating overcurrent protection and residual current detection (RCD). Easy access to the isolator must be provided at all times.
- The trailing, high voltage cable of the primary circuit must be heavily insulated and armoured against mechanical damage.
- Make sure that the cables associated with the low voltage, heavy current of the secondary circuit are adequate for the welding currents to be used. Also make sure that they are correctly and safely terminated.

- All cables and fittings must be frequently inspected for damage. Any damage to the cables and fittings must be corrected by a suitably qualified electrician.
- Make sure all equipment and the work are adequately earthed with conductors capable of carrying the heavy welding currents.
- Wear an arc-welding protective helmet or use a hand-held face shield. These contain special filter glasses that protect the user's face, eyes and head not only from the heat and glare of the weld pool but also from the ultraviolet radiation of the arc. It is because of the ultraviolet radiation of the welding arc that gas welding goggles are unsuitable when arc-welding.
- Adequate ventilation is required because of the fumes from the flux coating of the electrode and to keep the working environment cool. Fume extraction should be 'low level' so that the fumes are not drawn up past the face of the welder.
- The safety clothing recommended for gas welding is suitable for most arc-welding operations but additional protection will be required when working overhead.

Chemical joining and finishing

The main hazards associated with chemical joining and finishing can be summarized as follows:

- Inhalation of toxic and narcotic solvent fumes from chemical welding materials, adhesives and paints.
- Inhalation of toxic and narcotic solvent fumes from degreasing equipment.
- Fire. Many of the solvents and chemicals used in joining and finishing are highly flammable and even explosive. All areas where such chemicals are used must be adequately ventilated and they must be strictly *non-smoking* zones.
- Chemical splashes. Many of the chemical solutions associated with pre-process pickling, electroplating, anodizing and conversion coating contain strong acids. Strong caustic solutions are also used for degreasing. Chemical proof gloves, aprons, spats and face visors must be worn at all times when working with such chemicals.

Test your knowledge 4.19

1. Explain why goggles that are suitable for oxy-acetylene welding are unsuitable for electric arc-welding.

2. Explain why clear goggles should be worn:
   (a) when soft soldering;
   (b) when chipping the flux coating from a finished arc-weld.

3. Explain why ventilation is important when storing and using oxy-acetylene welding gases.

4. Explain why thick cables are needed for the welding circuit of an arc-welding installation.

5. Find out and describe the correct procedure for setting and lighting up oxy-acetylene equipment.

Key Notes 4.3

- The purpose of assembly is to combine together all the components and subassemblies necessary to make the finished product.
- There are three groups of joining processes used: mechanical, thermal and chemical.
- Mechanical joining techniques include threaded fasteners, keys, rivets and compression joints.
- Thermal joining techniques include soldering, brazing, fusion welding, friction welding, resistance welding and ultrasonic welding.
- In fusion welding the edges of the metals being joined are melted and allowed to run (fuse) together.
- Oxy-acetylene welding uses these gases burnt in a welding torch as the heat source. A separate filler rod is used to add metal to the joint.
- Electric arc-welding uses the heat of an electric arc (continuous, elongated spark) struck between an electrode and the work as the heat source. The temperature is much higher and the amount of heat energy available is much greater than for gas welding, therefore thicker metals can be joined. The electrode is also the filler material and is surrounded by a flux coating.
- In soldering and brazing only the filler material (solder or brazing spelter) becomes molten. The materials being joined are not melted. A flux is required.
- Chemical joining techniques include chemical welding and adhesive bonding. The adhesive used has to suit the materials being joined and the service conditions.
- Assembly processes are just as much subject to quality specifications as manufacturing processes. However, the quality assessment of assemblies is more concerned with finish and performance.
- Finishing processes treat the materials used during manufacture in order to provide decoration, retard corrosion and degradation, and protect the surfaces from wear.
- Fabrics may be self-colour or they may be dyed and printed to improve their appearance. They may be treated to make them water proof, crease resistant, flame resistant, drip dry, colourfast and stain resistant.
- Plastic products may be self-colour, painted, vacuum metallized, electroplated and polished.
- Wood products may be painted, varnished, lacquered, sealed and creosoted to improve the finish and render them rot resistant.
- Foodstuffs may be finished with icing, icing sugar, bun-wash, glaze and edible decorations.
- For higher temperature finishes such as for ovenwear, metal products can be vitreous enamelled.
- Metal components such as gas turbine blades are often ceramic coated to make them more resistant to the high temperature erosion.
- Metal components may be given protective and decorative coatings by such processes as electroplating, anodizing, galvanizing, etc.
- Metal components may be protected by conversion coatings, for example chromating and phosphating processes, prior to painting.
- Metals, glass and fabrics may be finished by a variety of heat treatment process after manufacture. These may harden, soften or stress relieve the products. In the case of fabrics they may be hot pressed to remove creases and to make pleats more permanent.
- Safety is as important to assembly and finishing processes as it is to manufacturing processes. This is particularly the case when welding is used.
- Welding equipment must never be used except by persons closely supervised while under instruction and persons who are fully trained in its use.
- When welding, the appropriate protective clothing must be worn and the appropriate protective equipment used. Gas welding goggles are unsuitable when electric arc-welding and arc-welding masks are unsuitable when gas welding.
- Oxygen and acetylene gases are highly dangerous and must be used and stored with the greatest care and in accordance with the practices recommended by the suppliers and the local authority fire officer.
- Electric arc-welding equipment must be inspected regularly by a qualified electrician.

Evidence indicator activity 4.3

1. Produce a log briefly recording your performance in practical exercises involving the assembly and finishing of one product as directed by your instructor, and which accords to the product specification.
2. Produce a supplementary record of involvement with one other product using contrasting materials.

Note: The log should record observation of your preparation and checking of machinery and equipment, components and materials, and the correct use of safety equipment, procedures and systems.

Supplementary evidence should be provided to cover those aspects of assembly and finishing not appropriate to the products being assembled and finished.

4.4 Quality assurance applied to manufactured products

Product quality has been introduced from time to time during the preceding elements. This element brings together many of the factors involved in manufacturing a product that is 'fit for purpose'. Quality assurance is achieved by adopting proven systems and procedures of inspection and record keeping. Typically, a company that is approved to ISO 9000 will provide goods and services whose quality can be assured.

4.4.1 Quality indicators

Quality indicators have already been introduced, and we saw that they could be *variables* or *attributes*. An indication of quality can be obtained by objective measurement (the component is or is not the correct size), or subjective assessment (the taste or feel is satisfactory in the opinion of the inspector). We saw that quality indicators are applied at critical control points during production. These quality indicators are essential in attaining the quality specification agreed with the customer.

4.4.2 Frequency of analysis

For 'one-off', prototype and small batch production every product is inspected. This is 100% sampling and inspection. It is also used for key components where a fault could cause a major environmental disaster or the loss of human life. Under these conditions, the high cost of 100% inspection would be small compared with the cost of failure in service. For the large volume production of many items, 100% inspection is too time consuming and expensive if carried out manually. Either fully automated 100% inspection is built into the production line or statistical sampling is used.

In statistical sampling, which we met in Section 3.3.5, samples are taken from each batch and the quality of the batch is determined by analysis of the sample. If the sample satisfies the criteria of the quality specification, then the whole batch is accepted. If the batch does not satisfy the quality specification the following lines of action may be taken:

- the batch is reinspected to confirm the original findings;
- the batch is sold at a reduced price to a less demanding market;
- the batch is reworked to make it acceptable;
- the batch is scrapped.

The most difficult decision when using statistical quality control is in selecting the most satisfactory sample size and frequency. If the sample size is too large, then the cost of inspection may be too great. If the sample size is too small, then the inspection results may give the wrong impression. The sample may be taken from the whole batch or, for very large batches and continuous production, samples may be taken on an hourly basis. Figure 4.54 shows the inspection criteria for our wooden carrying box which was first introduced to you in Fig. 3.2.

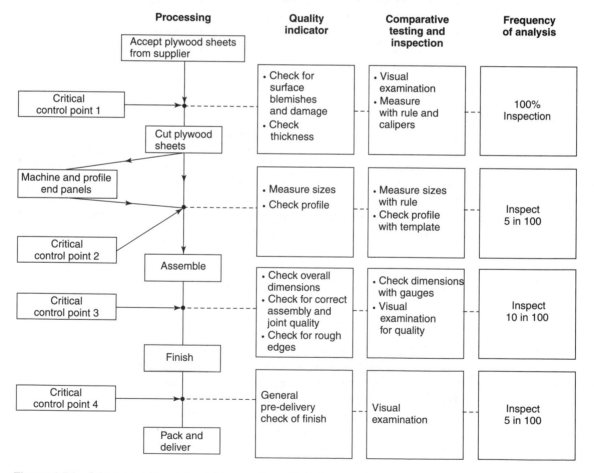

Figure 4.54 Critial control points and inspection criteria for wooden carrying box (Fig. 3.2).

4.4.3 Comparison of samples at critical control points

In Section 3.3.3 you were introduced to the idea that inspection is a comparative process. The quality indicator is compared with a known standard of acceptable quality.

- Dimensional measurements are compared with dimensional standards such as rulers, micrometers, calipers and gauges. These, in turn, are checked by comparing them with even more precise standards such as slip gauges.
- Paint and dye finishes are compared visually with colour charts.
- Food and drink products are assessed by highly skilled and experienced inspectors. These experts can compare the flavour of the current batch with their own memory.

4.4.4 Data recording formats

Statistical data can be recorded manually or it can be stored on a computer. One advantage of using a computer is that it can be loaded with a software package that will carry out the statistical analysis at the touch of a button and present the outcomes as either a graph, a chart or a spreadsheet. A computer can also handle very large numbers very quickly. Computers can also be used for much more comprehensive analysis including such factors as the amount of reworking that has had to be done, the amount of unreclaimable scrap produced, customer complaints, etc., and print out a weekly or monthly report.

However, for the moment we will consider a simple manually produced example. Table 4.8 shows the results of sampling a batch of wooden strips sawn from 25×50 mm planed timber. Each strip is to be 250 ± 0.5 mm long. A sample of 100 was taken from a batch of 1500 strips. The inspection results are then grouped to provide a *frequency distribution table* as shown in Table 4.9. Of the 100 strips sampled, 91 were within the specified limits of size and 9 were either oversize or undersize. That is, $(9 \div 100) \times 100 = 9\%$ were defective. This calculation supports the findings of Fig. 4.9.

You were introduced to *histograms* in Fig. 2.8, and this is a good way of showing a graphical representation of our frequency distribution. Such a graph is easier to interpret at a glance than a table of figures. Figure 4.55 shows the histogram for the results listed in Table 4.9.

Another way we can present the data for our wooden strips is to use a cumulative frequency distribution graph. First, however, we must produce a *cumulative frequency table*. If you look at the histogram you will see that the data recorded in Table 4.8 is the centre value of each column of the graph. The column spreads to each side of this value. The spread represents the *class limits* and the difference between the class limits is called the *class width*. Let's look again at Fig. 4.55. If we take the histogram column centred on 249.90 mm, you can see that its lower class limit is 249.85 mm and that its upper class limit is 249.95 mm. Since the difference is 249.95 mm−249.85 mm, the class width is 0.1 mm. We can now present our frequency distribution class limits as shown in Table 4.10.

Table 4.8 Results of inspection of sample of 100 wooden strips

Wood strip	Size (mm)	Wood strip	Size (mm)	Wood strip	Size (mm)	Wood strip	Size (mm)	Wood strip	Size (mm)
1	249.9	21	250.1	41	250.2	61	250.2	81	250.1
2	249.9	22	250.0	42	249.9	62	249.8	82	250.6
3	250.0	23	250.1	43	250.2	63	250.6	83	249.5
4	250.0	24	249.9	44	249.8	64	249.9	84	250.1
5	249.9	25	250.3	45	249.6	65	250.2	85	250.0
6	250.1	26	250.0	46	250.4	66	249.8	86	249.8
7	250.0	27	250.1	47	249.9	67	249.5	87	249.7
8	249.6	28	249.9	48	250.1	68	249.6	88	250.8
9	250.0	29	250.1	49	250.0	69	250.3	89	249.7
10	250.0	30	250.0	50	249.6	70	250.0	90	249.4
11	250.0	31	250.3	51	250.3	71	250.4	91	250.0
12	250.0	32	250.0	52	249.9	72	249.8	92	249.8
13	249.9	33	250.0	53	249.7	73	249.7	93	249.7
14	249.5	34	249.7	54	250.3	74	250.4	94	249.4
15	250.0	35	249.9	55	250.7	75	249.9	95	249.3
16	249.8	36	250.3	56	249.8	76	250.2	96	249.8
17	250.0	37	249.8	57	249.9	77	249.6	97	250.1
18	249.7	38	250.1	58	250.5	78	250.5	98	250.2
19	250.2	39	250.1	59	250.7	79	250.0	99	250.4
20	250.4	40	250.2	60	250.5	80	250.6	100	250.5

Table 4.9 Frequency distribution for wooden strips

Length in millimetres	Number of strips
249.3	1
249.4	2
249.5	3
249.6	5
249.7	7
249.8	10
249.9	13
250.0	19
250.1	11
250.2	8
250.3	6
250.4	5
250.5	4
250.6	3
250.7	2
250.8	1
	Total = 100

Having got over that bit of statistical mathematics reasonably painlessly, we will now draw up a *cumulative frequency table* using the upper class limits as shown in Table 4.11. This is not difficult for each of our previous frequency results we simply keep adding on the next result. This is what *cumulative* means, simply adding the next value to the previous one.

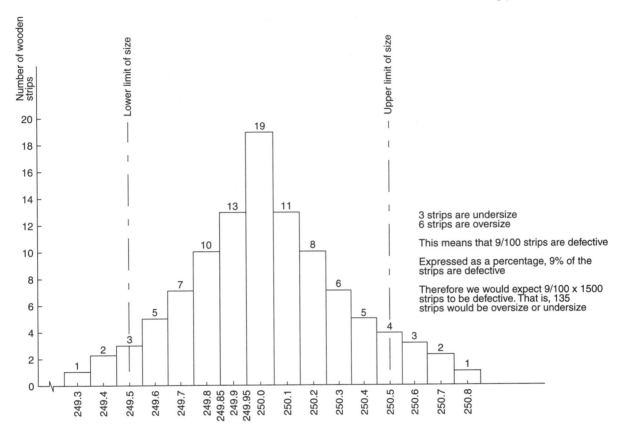

Figure 4.55 Histogram of the results listed in Table 4.8.

Again we can plot our results in the form of a graph so that we can see what is happening at a glance. The finished graph is shown in Fig. 4.56. The limits of size are also included on the graph. This shape of graph is called an *ogive* curve.

4.4.5 Defects in components and products

Defects are faults or flaws in products that do not comply with the quality specification. The four most important types of defect are as follows:

- critical;
- major;
- minor;
- incidental.

Critical defects

Critical defects are those likely to result in hazardous or dangerous conditions for anyone using or coming into contact with the product. An example of such a critical defect is a fault in the controls of an aircraft. Not only could

Table 4.10 Class frequency distribution for wooden strips

Length of strip in millimetres		Frequency
Lower class limit	Upper class limit	
249.25	249.35	1
249.35	249.45	2
249.45	249.55	3
249.55	249.65	5
249.65	249.75	7
249.75	249.85	10
249.85	249.95	13
249.95	250.05	19
250.05	250.15	11
250.15	250.25	8
250.25	250.35	6
250.35	250.45	5
250.45	250.55	4
250.55	250.65	3
250.65	250.75	2
250.75	250.85	1

Table 4.11 Cumulative frequency distribution for wooden strips

Upper class limit of length (mm)	Frequency	Cumulative frequency
249.35	1	1
249.45	2	$2 + 1 = 3$
249.55	3	$3 + 3 = 6$
249.65	5	$5 + 6 = 11$
249.75	7	$7 + 11 = 18$
249.85	10	$18 + 10 = 28$
249.95	13	$13 + 28 = 41$
250.05	19	$19 + 41 = 60$
250.15	11	$11 + 60 = 71$
250.25	8	$8 + 71 = 79$
250.35	6	$6 + 79 = 85$
250.45	5	$5 + 85 = 90$
250.55	4	$4 + 90 = 94$
250.65	3	$3 + 94 = 97$
250.75	2	$2 + 97 = 99$
250.85	1	$1 + 99 = 100$

this result in the loss of the aircraft, the crew and the passengers, but also the persons on the ground at the crash site.

Major defects

Major defects are those likely to result in failure or a reduction in the operating efficiency of a component or product, but not necessarily endangering life. An example of such a defect is the failure of the electric motor in a vacuum cleaner.

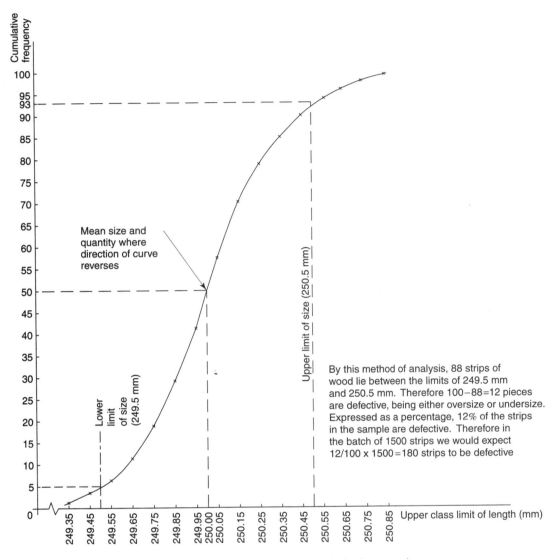

By this method of analysis, 88 strips of wood lie between the limits of 249.5 mm and 250.5 mm. Therefore 100−88=12 pieces are defective, being either oversize or undersize. Expressed as a percentage, 12% of the strips in the sample are defective. Therefore in the batch of 1500 strips we would expect 12/100 x 1500=180 strips to be defective

Figure 4.56   Cumulative frequency curve for wood strip sample (ogive curve).

Minor defects

Minor defects affect the saleability of a product but will only have little or no effect on its use. An example of such a defect is a loose button or a stain on a shirt.

Incidental defects

Incidental defects are those which have no effect on the use or function of a product. An example of such a fault is a small air bubble in a glass paperweight.

Defects and their causes

During manufacture defects will have many causes. Let's now look at some of the more common ones.

Operator errors

There are two main categories of operator error:

- Inadvertent errors are caused by carelessness, fatigue and inattention.
- Skill errors occur as the result of poor technique caused by inadequate training and supervision.

Inattention resulting in inadvertent error can occur at any time and is due to the operator not paying attention to the job in hand. The situation can be improved by 'fool proofing' the process, that is, designing the product or the process so that the parts can only be loaded into a machine or assembled the correct way round. This reduces the dependence on human attention. Job rotation within a work team, scheduled rest periods and designing the job to make it more interesting can all be used to improve operator attention and motivation.

Lack of skill or technique errors can occur when an operator lacks the skill or knowledge to perform a task correctly. This can not only result in the manufacture of products that do not meet the quality specification, but also can put the operator at risk of injury through the incorrect operation of equipment. This problem can be remedied by giving the operator extra training followed up by regular training reviews.

Process condition changes

Process conditions can change for a variety of reasons – for instance, a machine may be unable to hold its settings because of faulty maintenance. Precision measuring equipment can become inaccurate due to expansion and distortion if the temperature of the inspection area rises too much on a hot summer's day. Automated equipment is often extremely complex and a component failure or a virus in the control computer can cause a malfunction without necessarily causing the equipment to fail or shut down. An example of such a situation is when a soft drinks company has its bottle filling operation automatically controlled. A component failure or a computer virus will cause the process conditions to change and the control mechanism underfills the bottles such that the volume of liquid falls outside the acceptable tolerance limits.

Composition proportion changes

A common cause of the reduced quality of a product is brought about by changes in the composition of a product – for example, precast concrete beams are made to a specification in which the proportions of cement, aggregate and water are closely controlled. The proportions may change for a variety of reasons – for example, the metering and control equipment may get out of adjustment due to poor maintenance. Alternatively, an unscrupulous manufacturer might deliberately reduce the amount of cement and increase the amount of aggregate beyond the limits laid down to increase the profitability of the process. Such changes would reduce the quality of the product. Changes of process conditions or changes in composition proportions resulting from the malfunction of equipment must be rectified immediately. Planned, preventive maintenance should be organized to prevent such occurrences happening.

Material substitution

Material substitution in a product can affect product safety. Designers and materials engineers must carry out stringent and extensive testing before any change of material specification is permitted. Material substitution is usually made in an attempt to reduce costs or to tap more easily available sources. If this results in an unauthorized substitution of inferior or substandard materials, this will affect the quality of the product.

Test your knowledge 4.20

Table 4.12

Can no.	Volume (ml)	Can no.	Volume (ml)	Can no.	Volume (ml)	Can no.	Volume (ml)	Can no.	Volume (ml)
1	749	11	750	21	751	31	751	41	750
2	749	12	749	22	749	32	748	42	749
3	747	13	751	23	751	33	750	43	751
4	750	14	748	24	749	34	752	44	750
5	748	15	750	25	750	35	746	45	750
6	750	16	749	26	752	36	750	46	753
7	750	17	751	27	748	37	752	47	747
8	750	18	747	28	748	38	749	48	754
9	749	19	751	29	750	39	750	49	749
10	751	20	750	30	752	40	753	50	748

1. Given the data in Table 4.12, draw up:

   (a) a frequency distribution table;
   (b) a histogram for the results in the frequency distribution table;
   (c) a cumulative frequency table;
   (d) an 'ogive' cumulative frequency curve for the results in the frequency distribution table.

2. Classify the following defects as: critical, major, minor or incidental:

   (a) a blocked jet in a lawn sprinkler;
   (b) a knot in the wood of a window frame;
   (c) a fatigue crack in a gas turbine blade;
   (d) a slight run in the colour of a tea towel.

3. Give examples of your own choosing where the quality of a product would be reduced by:

   (a) a change in the process conditions;
   (b) a change in composition proportions;
   (c) the substitution of material.

4.4.6 Dealing with defective products and components

Defective products are those which are unacceptable because they do not conform to either the product specification or the quality specification – for instance, the door seals on a batch of washing machines may be defective causing flooding of the customer's kitchen. The remedial action that needs to be carried out on defective products depends upon the type of defect, the severity of the defect and the number of defective components involved. Let's now look at the procedures for dealing with defective products and components. Figure 4.57 shows a simple flow chart for planning remedial action.

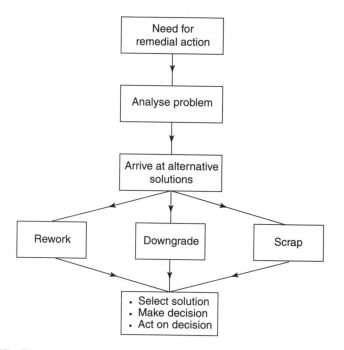

Figure 4.57   Planning remedial action.

Investigate cause

A quality deficiency in a component or product should be identified by the quality control system by inspection at the critical control points. If statistical process control is employed any deficiency will show up on the control chart, which will then be analysed by a quality control engineer in order to identify the defect and its probable cause. Once the defect and its cause have been identified as originating in a particular area of production or machine, a production engineer or a machine setter will be given the responsibility for taking remedial action to eliminate the cause of the defect.

The most likely causes of defects in manufactured goods could result from one or more the following:

- faulty or unsuitable tools and equipment;
- unsuitable materials;
- untrained or badly trained operators;
- unsuitable production methods;
- inadequate quality control system;
- poor product design.

Recommend solutions

The solutions recommended to eliminate the cause of the defect in the product will, of course, depend upon the product, the nature of the defect and the sector of the manufacturing industry concerned. It is easier and cheaper to correct a defect in a wheelbarrow than in a high performance military aircraft. Let's now look at some of the solutions available:

- Improve the tools and equipment being used.
- Introduce alternative materials or, possibly, materials of improved quality.
- Procure materials and components from more reliable sources.
- Improve training schedules and methods and retrain underskilled operators.
- Analyse and improved the production methods used so that quality standards are met without increased costs.
- Introduce efficient quality control systems, or improve the efficiency of existing systems.
- Improve the product design, but avoid overengineering so that it becomes uncompetitive with respect to cost.

Rework

Reworking is the rectification of defects on a product to meet the required specification. This is not always possible. Oversize products can be cut down in size but undersize products cannot always be rectified. If the wrong material has been used no rectification is possible (see Downgrade). Reworking is often expensive and the cost of reworking must considered carefully against the other alternatives.

Downgrade

A defective product is downgraded if it is usable but does not meet the required product specification, and is sold at a reduced price as 'substandard' or 'seconds'. This is often done with clothing and pottery, where 'seconds' are sold off to the public in the factory shop.

Scrap

Scrap is a defective product that is so far outside the specification that it cannot be reworked, used, or sold, and is only suitable for recycling. The scrap value of the material can be recovered and offset against the loss made on the batch.

Test your knowledge 4.21

1. Decide whether you would rework, downgrade, or scrap the following defective products giving reasons for your decisions:

    (a) A batch of M16 bolts that are tight in M16 nuts.
    (b) A batch of wing fixing bolts for a microlight aircraft made from defective material.
    (c) A china dinner plate with a fault in the glaze.
    (d) A batch of canned food that may have been contaminated.
    (e) A batch of neckties that are slightly short.

(f) Curtain material that has faded while in store.
(g) A batch of wooden window frames that are slightly oversize.

Key Notes 4.4

- Quality indicators are applied at critical control points and may be variables or attributes.
- Frequency of analysis is the sample size taken for inspection.
- Inspection is a comparative process where a feature of the product being manufactured is compared with a known standard that satisfies the quality specification.
- Formats for recording data may be manual, computer generated, tabular, or graphical.
- The recorded inspection data may be presented as: a frequency distribution table, a frequency distribution histogram, a cumulative frequency distribution table, or a cumulative frequency curve (an *ogive* curve).
- A *critical defect* can result in a catastrophic failure of a component or assembly resulting in loss of life or a major environmental disaster.
- A *major defect* will prevent the product operating correctly but will not involve injury or damage.
- A *minor defect* will have little or no effect on the performance of the product but will have lowered its quality so that it has to be disposed of as second grade at a reduced price.
- An *incidental defect* is so trivial that it will have no effect on the performance, appearance, or selling price of the product.
- Operator errors occur due to fatigue, boredom, or lack of training.
- Condition changes occur when equipment is unable to maintain its settings due to wear, inadequate maintenance, or computer viruses.
- Composition proportions occur when materials are being mixed or blended. The proportions may be changed accidentally or deliberately to increase the profitability of the product. In both cases the quality of the product is likely to suffer.
- Substitution of cheaper materials can be made to advantage. This should not be done without the consent of the designer, the materials engineer and the customer, as it may lead to a reduction in quality.
- Defective products may be *reworked* if this is economical or can be done without lowering the quality of the product. Defective clothing may be cut down and reworked as a size smaller providing the cost does not exceed the value of the cloth saved.
- Defective products may be *downgraded* if there is a ready market for the lower quality product. Second grade pottery and china is often sold off to the public in factory shops.
- *Scrap* is the name given to defective products that can neither be reworked nor downgraded. It is also the name given to the waste material left over when a shaped blank is cut from a standard rectangular sheet of the material.

Evidence indicator activity 4.4

1. Write a brief report for each of two products requiring different scales of production. The reports to include:

   - an outline of the relevant quality indicators and frequency analysis;
   - test and comparison data recorded in all the formats covered by those range of items;
   - a summary of the specific testing and comparison methods used;
   - a list of any defects with suggestions of possible causes and procedures for dealing with defective products, including consideration of all the procedures in the range.

Index